T0298651

DIGITAL GOVERNANCE

Digital Governance provides managers with a simple and jargon-free introduction to the impact that digital technology can have on the governance of their organisations. Digital technology is at the heart of any enterprise today, changing business processes and the way we work. But this technology is often used inefficiently, riskily or inappropriately. Worse perhaps, many organisational leaders fail to grasp the opportunities it offers and thus fail to "transform" their organisations through the use of technology.

This book provides an explanation of the basic issues around the opportunities and risks associated with digital technology. It describes the role that digital technology can play across organisations (and not just behind the locked doors of the IT department), giving boards and top management the insight to develop strategies for investing in and exploiting digital technology as well as arming them with the knowledge required to ask the right questions of specialists and to detect when the answers given are evasive or irrelevant.

International in its scope, this essential book covers the fundamental principles of digital governance such as leadership, capability, accountability for value creation and transparency of reporting, integrity and ethical behaviour.

Jeremy Swinfen Green MA, MBA, CMC, FIC has spent over 25 years advising organisations about digital technology and "human factors" – how people interact with technology. He has degrees from the University of Oxford and CASS Business School. He has also written *Cyber Security: An Introduction For Non-Technical Managers* (2015).

Stephen Daniels FMS, FIOR, FBCS, CITP has spent 35 years in digital governance, risk management, security, privacy, resilience and compliance. Whilst consulting to major organisations from BA to NATO, he has also authored over a dozen British and International standards in these disciplines.

DIGITAL GOVERNANCE

Leading and Thriving in a World of Fast-Changing Technologies

*Jeremy Swinfen Green
and Stephen Daniels*

Routledge
Taylor & Francis Group

LONDON AND NEW YORK

First published 2020
by Routledge
2 Park Square, Milton Park, Abingdon, Oxon OX14 4RN

and by Routledge
52 Vanderbilt Avenue, New York, NY 10017

Routledge is an imprint of the Taylor & Francis Group, an informa business

© 2020 Jeremy Swinfen Green and Stephen Daniels

The right of Jeremy Swinfen Green and Stephen Daniels to be identified as authors of this work has been asserted by them in accordance with sections 77 and 78 of the Copyright, Designs and Patents Act 1988.

British Library Cataloguing-in-Publication Data
A catalogue record for this book is available from the British Library

Library of Congress Cataloging-in-Publication Data
A catalog record has been requested for this book

ISBN: 978-0-367-07722-8 (hbk)
ISBN: 978-0-429-02237-1 (ebk)

Typeset in Joanna
by Deanta Global Publishing Services, Chennai, India

SD: To Krys, with thanks for her patience, support and encouragement

JSG: To Donna, Theo and Charlie, with apologies for the absent moments

CONTENTS

AUTHOR BIOGRAPHIES

Jeremy Swinfen Green MA, MBA, CMC, FIC
Jeremy Swinfen Green has spent over 25 years advising businesses about their use of digital technology. His focus is on how people (workers and customers) respond to and interact with technology: so-called "human factors". He has worked in strategy, marketing, customer experience, innovation, knowledge management and cyber security. He has a first degree in Natural Science from the University of Oxford and an MBA from CASS Business School.

Steve Daniels FMS, FIOR, FBCS, CITP
Steve Daniels has spent the last 35 years in digital governance, risk management, security, privacy, resilience and compliance. Whilst working for a multitude of major national, international and global organisations – from BA to NATO and the FCO to HP – he has also found time to author more than a dozen UK and International standards in these disciplines.

1

INTRODUCING DIGITAL GOVERNANCE

Summary

Digital governance involves the creation and monitoring of policies for invest-ments in and use of digital technology across an organisation. The increasing power of computers combined with lower prices means that digital technology is now ubiquitous. As well as providing the communication and data process-ing traditionally managed by IT departments, digital technology impacts on the way people work, the efficiency of factories, the ability to sell to consumers, and the strategic management of organisations. Ultimately it is concerned with the long-term strategic issue of how organisations can thrive during a time of rapid technological change.

   With this extended impact come extended risks and ever-growing oppor-tunities. Digital technology is now far too important to be governed by one element of an organisation. It is a strategic issue that needs to be addressed at the very highest level, by the leaders of organisations. This is the manifesto that this book sets out: the boards of organisations need to take responsibility for setting the digital technology agenda, identifying the opportunities it brings and ensuring the associated risks are appropriately managed. It is only the board, with their viewpoint extending across the whole organisation, that can do this effectively.

A digital governance manifesto

We live in turbulent times. The scale and nature of emerging risks around the use of digital technology in businesses and other organisations, including privacy, security and mental well-being, represent a real threat to organisational survival, let alone success.

At the same time, public expectations are on the increase, often out of line with what is possible for organisations to deliver. And alongside these expectations, new regulatory and legislative obligations, often transnational in effect, are also emerging.

In addition, there is a growing perception that the leaders of many organisations have failed to make the grade in terms of governance. Shareholder revolts, which have become increasingly common, are a symptom of this, often triggered by perceived or actual under-performance.

Why is this happening? We believe it is because we are living through a digital revolution that has been progressing at a breakneck pace for the last 20 years, a revolution that sometimes seems as if it is leaderless, rudderless and out of control.

This revolution is being driven by a number of new realities. Computers are ever-cheaper with more processing power. Machines continue to shrink, making them easier to embed in other machines. The interfaces between computers and humans are becoming more effective, meaning that some computers can actually be considered for physical embedding in people. Software development costs are falling, facilitating innovation. Enhanced sensors and extensive connectivity provide improved data collection, sharing, access and tracking. And new business services offering the opportunity for small organisations to rent massively powerful computing by the hour create an asymmetric environment where small and agile companies can now compete with international corporations without the need for huge investments in technology.

A common and widespread reaction of many leaders to this seems to be to treat digital activities as either a veneer on the traditional organisation, or a side issue to be delegated to an IT director or even some sort of stand-alone element of the organisation such as a digital division.

This treatment is then used to justify limited engagement with digital technology at board level, and to validate a perception that there is little or no need for the board to learn new skills to manage it.

We believe this attitude to be a critical mistake. Digital technology is of fundamental strategic importance to any organisation. And so, it follows that digital governance is also of fundamental importance to the organisation – and to its leaders.

This is the manifesto that we outline in this book: governing bodies must engage closely with digital technology and treat it as central to their governance activities, just as they treat issues such as cash flow, talent management and corporate reputation.

Understanding how digital technology should be governed is no easy task, however. For a start, there are many digital untruths that cloud judgements and decision making. You will probably have heard, and may even have believed, a number of widely accepted statements such as:

- *Everyone shops online these days.* Even if this were true, which it isn't, 80% of UK retail is still offline (1) and the figure for the USA is nearer 90% (2).
- *No one watches TV anymore.* UK adults watch 3 hours and 20 minutes of television a day on a TV set (3) and even 18- to 34-year-olds watch over 2 hours a day on average, a third of which is online, e.g. Netflix and BBC iPlayer (4).
- *Direct mail doesn't work anymore.* The return on direct mail is still an average of £3.22 for every £1 spent (5).
- *You don't need anti-virus software on an Apple computer.* Apple admitted that its devices are not immune from malicious software back in 2012 (6).
- *Everyone has a smartphone these days.* Not quite: around 20% of people in the UK and 45% of people in Japan don't own a smartphone (7) and globally the figure is nearer 60% (8).

Even if you avoid believing untruths, deriving an appropriate strategy and framework for digital governance is by no means an easy thing to do. There are numerous considerations, often conflicting. Trade-offs will undoubtedly be the order of the day and collaborations, for instance within trade associations or value chains, will be key – even if that means you are sitting down with the enemy.

Hard decisions will be needed. How can we decide between what is possible to do and what is worthwhile doing? Should we be at the bleeding edge of technology or should we be followers? Should we follow a stepped approach to digital transformation, or do we need a big bang? Should we be loud and proud about digital investments (which might go wrong) or quiet and reserved? The list is a long one.

Certainly, the only things about digital governance that there's agreement on are: there's no "one size fits all" answer; best practice is still being identified and codified in standards; and there's a dearth of success stories out there.

SOME DEFINITIONS

Throughout this book we have used "governing body" and "the board" interchangeably. We have done so as we believe that the substance of this book is applicable to all organisations. Similarly, if we have used "company" elsewhere, that is not intended to limit our message to commercial organisations.

Our use of "top management" is a reference to the executive community as against the board of directors or governing body to which it is normally subservient, as the duties of each are materially different, remembering of course that in many organisations some board-level directors will have an executive function.

The scope of digital governance

All organisations use digital technology to communicate with their stakeholders, to manage and transfer money, and to enable their operational processes. Technology underpins the money and people in organisations. It needs to be a key focus of organisational leaders and of top management reporting on its executive activity to the governing body.

The Institute of Charters Accountants in England and Wales (ICAEW) (9) defines corporate governance as follows: "Corporate governance is about what the board of a company does and how it sets the values of the company, and it is to be distinguished from the day to day operational management of the company by full-time executives". (10) This definition neatly summarises the split of responsibilities and validates that digital governance is therefore merely one element of corporate governance – but an ever-more fundamentally important one.

It should also be noted though that digital governance and IT governance are not the same. IT governance can be defined as the processes that ensure the effective and efficient use of IT in enabling an organisation to achieve its goals. It is very much the responsibility of the people who head up the IT function in organisations and its concerns are often short-term.

Digital governance is wider than this. Yes, it encompasses IT and the processing of data and information. But it is also concerned with the way that data can be used to make strategic decisions. It is concerned with the way that people's work and home lives are affected by the way they use digital technology during work hours. It is concerned with the way that factory machines can be designed and maintained with the help of digital technology such as augmented reality and digital twins. Ultimately it is concerned with the long-term strategic issue of how organisations can thrive during a time of rapid technological change.

As with corporate governance more generally, digital governance requires that we consider a number of elements when thinking about how digital governance needs to be established:

- **Corporate objectives and values**. Governance starts with the mission, values and goals of the organisation and how they are achieved. In the case of digital governance, there is a need to understand how the use of existing or emerging technology by the organisation or other organisations can enable or hinder the mission, values and goals.
- **The organisational context**. There are many internal and external issues that will affect the ways that an organisation can use, or will be affected by, digital technology. These will include the policies and processes that define how the organisation operates, for instance, decisions that have been made about automating a factory and the consequent effect this has had on employees, profitability and competitiveness.
- **Stakeholders**. The principal stakeholders of any organisation are its owners but there are other stakeholders who should be considered by

the board, including employees, customers, regulators, suppliers, competitors and society in general.

- **The market**. Proper governance must consider the markets that an organisation operates in (or could operate in) and how the individuals (human or organisational) who make up those markets are, or could be, affected by technology. This then means considering where digital technology is headed as well as maintaining an awareness of the actions and intentions of any competitors just as much as your own plans.

- **Compliance requirements**. Any legislation (such as the Equalities Act 2010 or the Data Protection Act 2018), regulations (such as the UK advertising codes laid down by the Advertising Standards Authority) and standards (such as ISO 9001, the Quality Management standard and ISO27001, the information security standard) that the organisation has to, or chooses to, abide by.

- **The board**. It is important for any board to have unbiased self-knowledge about how effective they are in terms of digital governance. Their knowledge of technology, the way they behave around it, and their opinions about its importance and potential will not only influence the culture of the wider organisation but also the ability of the board to ensure appropriate digital governance. To be kept honest and not to become complacent, governing bodies really ought to obtain independent, third-party feedback on the effectiveness of their digital governance. At least one major UK bank has obtained this, within a wider feedback and input process, via a Technology Advisory board, reporting to the Chief Technology Officer who sits on the bank's board.

DIGITAL TECHNOLOGY AND THE WATES PRINCIPLES OF CORPORATE GOVERNANCE

The Wates corporate governance principles (11) are most relevant for very large private organisations but make a useful framework for other companies. They include several principles of governance that have relevance to digital technology.

1. **Purpose and leadership:** An effective board "develops and promotes the purpose of a company, and ensures that its values, strategy and

culture align with that purpose". This includes developing a business model that generates long-term sustainable value. Our feeling is that it will be very difficult to do this without an in-depth consideration of technology and the opportunities and threats it brings.

2. **Board composition:** An effective board needs "a balance of skills, backgrounds, experience and knowledge". In the past the focus here has been on a knowledge of how profits are made and how the right team to make those profits is assembled. Given its increasing importance, a knowledge of technology surely has to be included in an effective board.

3. **Director responsibilities:** In line with the UK Companies Act 2006, directors on organisational boards have a duty to "promote the success of the company for the benefit of its members (shareholders) as a whole". They are accountable for the way the organisation operates, and accountability requires integrity of information. This means that boards need to ensure that included in the information they receive, there should be sufficient information about market trends including technology trends: the ability to scan and analyse the likely effect of technology on the short-term and mid-term future of the company is crucial.

4. **Opportunity and risk:** Boards should "promote the long-term sustainable success of the company by identifying opportunities to create and preserve value and establishing oversight for the identification and mitigation of risks". Technology brings with it many opportunities, but also many risks, not least if the opportunities are ignored as many large organisations have discovered. Blockbuster, Polaroid and Kodak are three frequently cited examples (12).

5. **Stakeholder relationships:** "(E)ffective stakeholder relationships aligned to the company's purpose" need to be fostered; and *one* particularly important stakeholder group is the workforce. The effect of technology on the willingness and ability of employees to work productively should be a priority for consideration by boards.

Building on this and reflecting continued concerns about the quality of organisational governance, back in 2013, the British Standards Institution (BSI) facilitated the drafting of BS13500, which specifies the governance processes required in more detail. This standard has led to the development of an International Standard, ISO37000.

Despite the fact that all organisations of any size are completely reliant on digital technology, only 5% of non-technology companies have digital expertise on their board. Given its strategic importance, a failure to govern digital technology will inevitably put an organisation at risk.

Even for a business trading in the physical world, one making car parts, for instance, or packing food, digital technology is an essential component of operations. And that's because digital technology enables the collecting, sharing and using of data and information – organisational assets that are just as important as raw materials, machinery and people. That information can very quickly become valuable intellectual copyright in its own right.

These assets can be found everywhere in any organisation and they affect the ability of every part of an organisation to perform well. For instance:

- Digital technology (obviously) underpins any digital transformation initiatives.
- Many regulations, including those relating to privacy, human resources, safety at work, finance and marketing involve the way organisations can, or must, use information.
- The analysis of data and information, increasingly powerful with the use of artificial intelligence and other software tools, enables effective strategic plans to be developed and followed.
- Customer trust (or distrust) is closely linked to the way that the digital technology used protects (or otherwise) privacy, prevents fraud, enables good customer service and projects (or not) a positive organisation reputation.
- The ability to deal with third-party organisations around the world, including suppliers and other business partners, is also closely linked to the way that digital technology allows communication and interconnectedness.
- Effective technology including good cyber security allows organisations to take risks that otherwise they would be unable to consider.
- Digital technology affects the skills that organisations need from their workforce, as well as enabling employees to share knowledge and grow skills.

Overall good digital governance is associated with better business results. But it isn't easy to achieve. And that's why it must be handled as a strategic issue, appropriate for board-level consideration.

The role of the governing body

Digital governance starts and ends with the board of the organisation. It is these organisational leaders who must set the direction of their organisation's relationship with digital technology for others to execute – they provide guidance on how the organisation makes decisions about its use and how it reacts to the change (and, sometimes, chaos) that technological developments can bring.

What does this mean in practice? Leaders will set the direction by identifying and communicating three things:

1. **The goals** of the organisation's investment in digital technology. For instance, are there opportunities to differentiate their organisation through technology? Should technology be used to secure the organisation's IP? Can efficiency be increased, or costs reduced, through technology? In short, what are the strategic opportunities that technology brings now and in the near-to-medium future?
2. **The prudence** that the organisation should show when investing in and using technology. How much risk are they willing to accept when using and investing in technology? How much capital and revenue are they prepared to invest in it? Cyber security is an obvious area of consideration but, as we shall see, there are many others.
3. **The ethical stance** organisations should take if considering technology. In particular, they must decide what uses of digital technology they feel are inappropriate for their organisation from an ethical (and reputational) standpoint. How they treat the privacy of their customers is one important area and how they treat their employees is another. The UN's Sustainable Development Goals (13) make a useful framework here.

Setting the direction is only part of the task of organisational leaders though. They also have to see that the direction they have set is followed and they then need to act upon the results.

It is therefore helpful to think about the board's role as having three main areas: accountability, direction and control. In each of these areas you will be working with your top management to:

(1) acknowledge your ultimate accountability
(2) set the direction you want followed

(3) ensure the implementation of appropriate controls
(4) accept your accountability again for the success or otherwise of the collective leadership given.

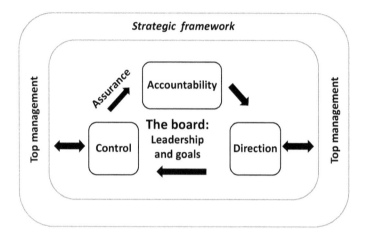

Figure 1.1 The relationship between an organisation's board and its top management.

Direction and Control are supported by a strategic framework that the board should validate even if they don't set it (Figure 1.1) (see Chapter 2).

- **Accountability**. The board of an organisation is accountable to the owners of the organisation and other stakeholders in respect of the investments made in, and uses of, digital technology. They should acknowledge this and at the same time define a clear purpose for technology that can support the wider organisational goals around the quality of products or services, growth of the organisation over time, and the financial strength of the organisation.
- **Direction**. Leaders should establish what their expectations are around the goals they have set and review whether any objectives, policies, processes etc. that management have put in place to support goals are appropriate.
- **Control**. Leaders need to ensure that their leadership team has put controls in place to ensure that if the direction they set for technology is not followed, remedial action can be taken. Controls will include requiring senior managers to provide sufficient information, informing them if

organisational performance such as growth or quality is inadequate, or requiring changes to major policies that are proving to be flawed.

While it is not appropriate for board members to become involved in day-to-day managerial matters, especially the details of digital technologies, they may from time to time be called on to support the Chief Executive to intervene with management and influential stakeholders including important suppliers and others, for instance regulatory authorities, if the organisation is performing poorly or being obstructed in a particular area.

They should also set the rules around the sort of decisions that the board, rather than top management, should take. For instance, certain investments in technology such as a new customer database or the automation of factory processes may be some of the largest an organisation may make and it is therefore entirely appropriate for the board to be closely involved and to take the final decision whether to adopt and to declare the limitations of any decision to do so.

- **Accountability (again)**. The Accountability-Direction-Control cycle is just that – a cycle, starting and ending with the accountability of the board. Leaders, as well as defining the purpose of technology within their organisation, are accountable for its proper use through their oversight of systems that affect or are affected by technology, their review of the direction they have set, and their actions, delegations of authority and instructions to others, to control the way and the degree to which digital technology helps to deliver the goals of the organisation.

The Accountability-Direction-Control cycle can only be effective if there is a strategic framework in place that supports and enables the successful use of digital technology by defining the fundamental principles, policies and practices that your organisation has committed itself to following. We examine this in Chapter 2.

About this book

The breadth of digital governance has long been one of the biggest hurdles to its successful implementation. In this book, we have sought to provide a thorough introduction to each of the topics that make up this vital domain.

But it is impossible to cover the whole domain in detail without creating the equivalent of an encyclopaedia.

In writing this book, we have drawn upon our own, real-world, experiences, as well as the knowledge shared by others. Our aim is to explain what digital technology means for organisations, and to describe how it needs to be governed by the leaders of organisations. And while this book is primarily aimed at governing bodies, and people who aspire to join them, we hope it will be valuable to anyone with an interest in how their organisation should be managed, whether they are directors, top management, Accountable Officers, members of specialist project teams, or "ordinary" employees, all of whom will be using, and be affected by, digital technology while they are at work.

Each of the subjects we cover could, on their own, fill a book. Indeed, you will find many books on each (14). We hope you find this overview helpful, before perhaps you assign responsibility to colleagues for each of the areas that are relevant to your organisation. They can then obtain and read those other books while you, with the aid of this one, ensure that they are travelling in the right direction!

References

1. ONS. (2019). *Internet Sales as a Percentage of Total Retail Sales (Ratio) (%) – Office for National Statistics.* [online] Available at: www.ons.gov.uk/businessindustryandtrade/retailindustry/timeseries/j4mc/drsi [Accessed 24 May 2019].

2. Statista. (2019). *United States: e-Commerce Share of Retail Sales 2021 | Statistic.* [online] Available at: www.statista.com/statistics/379112/e-commerce-share-of-retail-sales-in-us/ [Accessed 24 May 2019].

3. Thinkbox. (2019). *2018 TV Viewing Report.* [online] Available at: www.thinkbox.tv/Research/Nickable-Charts/TV-viewing-and-audiences/Monthly-Report#download [Accessed 24 May 2019].

4. Marketing Charts. (2019). *The State of Traditional TV: Updated with Q3 2018 Data – Marketing Charts.* [online] Available at: www.marketingcharts.com/featured-105414 [Accessed 24 May 2019].

5. Royal Mail. (2019). *Building a Business Case for Direct Mail | Royal Mail Group Ltd.* [online] Available at: www.royalmail.com/business/insights/how-to-guides/building-a-business-case [Accessed 24 May 2019].

6. The Register. (2019). *Even Apples Sometimes Have Worms in Them, Admits Cupertino*. [online] Available at: www.theregister.co.uk/2012/06/26/apple_computer_virus/ [Accessed 24 May 2019].

7. Newzoo. (2019). *Top Countries/Markets by Smartphone Penetration & Users | Newzoo*. [online] Available at: newzoo.com/insights/rankings/top-50-countries-by-smartphone-penetration-and-users/ [Accessed 24 May 2019].

8. quoracreative.com. (2019). *101 Mobile Marketing Statistics and Trends for 2019*. [online] Available at: quoracreative.com/article/mobile-marketing-statistics [Accessed 24 May 2019].

9. www.icaew.com/

10. ICAEW. (2019). *What Is Corporate Governance?* [online] Available at: www.icaew.com/technical/corporate-governance/principles/principles-articles/does-corporate-governance-matter [Accessed 24 May 2019].

11. FRC. (2019). [online] Available at: www.frc.org.uk/getattachment/31dfb844-6d4b-4093-9bfe-19cee2c29cda/Wates-Corporate-Governance-Principles-for-LPC-Dec-2018.pdf [Accessed 24 May 2019].

12. Collective Campus. (2019). *10 Companies that Failed to Innovate, Resulting in Business Failure*. [online] Available at: www.collectivecampus.com.au/blog/10-companies-that-were-too-slow-to-respond-to-change [Accessed 24 May 2019].

13. United Nations. (2019). *Transforming Our World: The 2030 Agenda for Sustainable Development. Sustainable Development Knowledge Platform*. [online] Available at: https://sustainabledevelopment.un.org/post2015/transformingourworld [Accessed 24 May 2019].

14. For the chapter on cyber security we would naturally recommend: Swinfen Green, J. (2015). *Cyber Security: An Introduction for Non-technical Managers*. London: Gower.

2

DIGITAL GOVERNANCE STRATEGY

Summary

At the heart of digital governance are the strategic questions: Where are we now, where do we want to be and how are we going to get there? These are the questions that leaders need to address if they are to guide their organisations through today's rapidly changing technology landscape. And given the extent of digital technology within almost all organisations, these are hard questions to answer.

A start can be made by auditing your organisation's maturity in its use of digital technology. This, combined with insights into consumers and competitors, can help you develop a vision. But moving towards this vision is inevitably difficult, especially as the technology landscape is constantly shifting. However, there are a number of basic principles encompassing issues such as risk management, customer focus and information management, underpinned by situational awareness and continuous improvement, that can enable success. It is, though, the task of organisational leaders to achieve this success. Ultimately digital technology is their responsibility.

It's probably not a wonderful idea to have a digital strategy for its own sake. Asking yourself "How can we transform our organisation into a digital organisation?" isn't a particularly fruitful, or indeed meaningful, question unless it is truly integrated into the other strategies you already have and into the business's DNA. Being digital should *not*, of itself, be an organisational goal, unless of course you are in one of those few industries that are being transformed totally and utterly by technology, such as media, communications and IT. Even then, although the scale may be bigger, it is still a means to an end and not an end in itself.

Instead of thinking about how you can bolt more digital technology onto your organisation, you should think about digitising your business strategy. How can you use technology to achieve your goals in a world where you, your customers and your suppliers are increasingly using computers (in everything from cars and phones to security cameras and cookers, and even tarmac, clothes, supermarket shelves and sewers) to do everything, everywhere and all the time?

The danger of having a separate digital strategy is that it will inevitably end up focussing on the enabling technology, rather than on your business. Digital technology isn't somehow separate from the rest of your organisation. And therefore, any strategic thoughts you may have about it obviously need to take account of, and be subjected to, your overall business strategy.

In this chapter we take a look at the main strategic issues that you will need to consider when ensuring appropriate digital governance.

A strategy for digital technology

Any business strategy can essentially be expressed, albeit simplistically, in answers to these three fundamental questions:

1.   Where are we now?
2.   Where do we want to be?
3.   How are we going to get there?

Answering these questions is one of the critical tasks of organisational leaders. In terms of digital technology strategy, the first question, "Where are we now?", involves understanding the technology status of an organisation. This will involve an audit of what technology is currently being used (and how up to date it is), how it is being used and why it is being used that way. This is a relatively simple question to answer in theory. In practice, though, due to the amount and variety of digital technology in organisations, it may be hard to answer. However, as a technology audit is a very necessary part of remaining cyber secure, your organisation may well have addressed this particular problem.

The second question, "Where do we want to be?", is harder. Understanding this requires knowledge of the potential for your organisation. And it may well be hard to see what that potential is, given that technology is changing so rapidly. A vision can only be developed as a result of an examination of consumer trends, competitor activities and some form of horizon scanning for technology developments.

The third question, "How are we going to get there?", is in some ways the simplest to answer, at least for organisational leaders who do not personally need to dive into the weeds of operational planning. Setting policies and guidelines around what is acceptable practice and what is not, in terms of risk, ethics, return on investment (ROI) and reporting will be second nature to most organisational leaders, even if they have not set these around technology before.

Agreeing on the current situation

Where are we now? A key function of the governing body is to ensure the auditing of the current status of the organisation in relation to digital technology. The adoption of digital technology by any organisation cannot happen in one go: it's a journey. Most organisations will have started on that journey (do you know any organisations that don't use a computer or a smartphone for at least something?). But it's by no means true that most organisations are getting towards the end.

Auditing where you are in this journey and measuring the maturity of your organisation's thinking about technology is the first step to creating a structured plan for digitisation. You can start by thinking where your organisation is on this spectrum:

Essential. The organisation uses the digital technology it needs to – a mobile phone for communicating, a computer for preparing documents, a simple website for signalling products and services. An IT function is the main driver and what it does do it does reasonably well but the need for its extension is at least recognised. Leaders are not really engaged with technology.

Extension. The organisation acknowledges that there is more to do and is actively seeking to identify the areas where, and the means by which, it could improve efficiency through technology. This might include developing a transactional website or using accounting software. Leaders are still not very engaged but may ask questions about how technology can help streamline certain processes.

Experimentation. The leadership of the organisation realises that there is more to digital technology than small increases in efficiency. This is a strategic issue where step changes in efficiency can be achieved, or where products and services can be transformed. One or two areas are chosen for trials that look to achieve the sort of step changes that are possible. Leaders ask to be kept informed.

Enterprise-wide. There is acceptance that a full reworking of the organisation's technology strategy is needed and active personal engagement at board level. Instructions are given to top management by the governing body to plan for increased use of digital technology across the organisation and to report on success and failure to the board; appropriate governance is under development.

Evolving. Robust processes for the continued identification and adoption of new digital technology are in place. The board is fully engaged with setting the direction of travel and monitoring process; they realise that this is a never-ending journey.

Of course, an organisation will not sit in a single position on such a five-step spectrum such as this. Technology and business have a far more complex relationship than that. But you can probably place your organisation somewhere on this spectrum, perhaps with some parts at stage 2 and other parts at stage 3 or 4. At the very least, considering your position on this spectrum (or others like it) is a useful exercise, and possibly a salutary one, especially as, in our view, many organisations, even very large ones, are in reality only at stage 2 or 3.

As well as auditing the technology status of an organisation, it is likely to be useful to audit its digital governance maturity. To do that a simple model can be constructed which will enable you to measure your current governance status in a number of areas.

In the model proposed on page 19, we have identified a number of governance areas along the top axis, with the five stages of technology adoption that we identified just now, which are shown on the left. In the cells of the matrix, we have suggested some evidence that might indicate an organisation has reached a particular level of governance maturity (Table 2.1).

The benefit of these two assessments will of course be that you will be able to identify if the digital strategy and digital governance of the organisation are aligned or whether the former (strategy) is being achieved, whether a little or a lot, despite rather than due to the latter (governance). If that was to be so, then shame on that board!

Developing a vision

It's all very well having a set of principles to follow. But how do they help with the central task of leading an organisation by developing its overarching vision, mission and strategic goals?

ARTIFICIAL INTELLIGENCE AND ORGANISATIONAL STRATEGY

Can machines really help you develop an effective strategy? Quid, an artificial intelligence (AI) company based in the USA and the UK, claim they can. "Human intuition at super human scale!" is their strapline.

According to Quid, its technology "searches, analyses and visualises the world's collective intelligence to help answer strategic questions". It does this by scouring the internet looking at company reports, news and social media.

But it inevitably has limitations. Quid, like any artificial intelligence system, will only collect what it is programmed and able to collect. Much of the information on the internet isn't available to robot searches. And of course, the internet isn't the only place that human information is stored: a lot is stored in people's heads and exchanged as they talk and write.

And the analysis that an artificial intelligence can conduct is limited too. It can curate information very efficiently by collecting, prioritising and structuring it, according to its own rules. But it may not find the hard-to-see connections that humans can (for instance the link between a strawberry picker and an aeroplane – see later in this chapter). And analysing the probabilities of highly illogical human behaviour may also provide substantial difficulties.

A tool like Quid can never be a substitute for thought. But then it doesn't claim to be. Rather it is an aid to thinking. It "draws connections between big ideas, giving your brain more power". Like much technology, Quid doesn't substitute for humans. It enhances them.

Table 2.1 A maturity model for digital technology governance

Maturity level	Board's personal openness to technology	Board-level knowledge of existing and emerging technology	Board-level involvement in setting technology policies and guidelines	Board-level involvement in monitoring and reporting
Essential	Board members show little interest in technology beyond entertainment, shopping and remote working	Technology is not raised as a strategic issue at board level	Board has no interest in technology policies and guidelines	Board only receives reports about major technology incidents
Extension	Most board members actively use technology e.g. mobile apps to increase work efficiency	Board receives occasional briefs on technology from senior managers	Board requires technology policies and guidelines to exist	Board receives regular updates about operational technology issues, e.g. IT spend and cyber security
Experimentation	Most board members are aware of and interested in technology change, as demonstrated by reading and research	Board receives occasional briefs from technology experts, generally based on topics in the news	Policies and procedures are signed off by the board; risk management processes adopted	Board engages in pro-active questioning of top management about technology status and the use of emerging technology
Enterprise-wide	Most board members demonstrate good practise in their own use of technology, e.g. in cyber security	Board receives regular briefs from technology experts. Venture capital and/or academic partnerships used to increase knowledge	Proactive involvement by the board in setting policies and procedures; roadmap for technology implementation established	Board reviews ROI and other key performance indicators (KPIs) of technology; taking an active interest in how failures can be turned into successes
Evolving	Some board members are enthusiastic early adopters of emerging technology and are eager to share their experience with peers	Board members actively expose themselves to ideas about emerging technology in business and can explain these issues to their peers	Strategy, policies and procedures regularly reviewed in the light of emerging technology	Board members champion the trial of untested technologies

A big challenge for organisational leaders is to understand when technology can be used to replace humans, perhaps taking them away from dull and dangerous jobs, and when (and how) it can be used to enable humans to perform better and add greater value. This understanding is essential for the development of a technology vision that maximises the opportunities both of technology and of humans.

How will we get there?

It is of course the board's role, supported by its top management and relevant subject matter experts (often these days referred to as Chief Officers) and especially its Chief Digital Officer (if it even has one), to set out the roadmap through which it will deliver its digital ambitions as defined in its digital strategy. As discussed in the introduction to this chapter, your strategy (and its associated roadmap and implementation plan), will also need to emphasise how the important role of digital governance will be implemented and delivered. This is not any sort of add on to the digital strategy, not least as it is fundamental to the control of your digital investments and to extracting the maximum services delivery benefit from them.

A pragmatic approach to digital governance

Alongside an overall acceptance of the board's accountability with regard to the use of digital technology (see Chapter 1), there is also a need for a practical approach to making decisions around its use.

This does not mean that your board should define the detail of the elements of how digital technology is managed across the organisation, except perhaps where major decisions have to be made. Instead, you need to understand whether the system you have in place for the governance of digital technology is effective in delivering against the organisational goals you have set. In other words, they should be confident that technology is playing its full part in supporting the success of the organisation.

Strategic principles

To achieve this, we believe there to be a strategic framework of twelve principles, supported by policies, rules and processes that can be adopted. These principles are:

1. **Command of the subject:** Board members must have adequate knowledge to lead and oversee investments in, and the actual uses being made of, digital technology.
2. **Accountable officer:** A specific person or team must be appointed to manage strategic digital technology issues.
3. **Situational awareness:** The board should ensure that strategic decisions taken around digital technology are being informed by the current and likely future situation.
4. **Ethics, policies and principles:** The approach of the organisation to the use of digital technology must reflect its wider approach to doing business.
5. **Information management:** There must be an understanding at board level of the need to gather, evaluate and curate the information and intellectual property needed to operate the organisation efficiently.
6. **Risk management:** The board must ensure that risks caused by the use of digital technology are identified and managed in accordance with the risk appetite agreed by them for the organisation.
7. **Appropriate technology use:** The board must ensure that the investments made in digital technology support the organisation's purpose.
8. **Customer experience:** The board must be satisfied that the end-to-end experience of customers, including digital and non-digital elements, is appropriate and on a path to eventual optimisation.
9. **Community engagement:** The board must encourage top management to collaborate with the widest range of stakeholders to ensure the most effective use of digital technology.
10. **Resilience planning:** The board must ensure that processes are in place to strengthen organisation resilience in the event of failures caused by digital technology.
11. **Continuous improvement:** The board must periodically review its own performance in overseeing digital technology across the organisation effectively.
12. **Flexibility:** The board must ensure the organisation is able to adopt new technology, abandon old technology and reject failed technology trials as appropriate.

We will explore each of these principles in a bit more detail later in this chapter.

Achieving good governance

The board must ensure that digital technology is always treated as *a servant of the business*. Investments in hardware and software should be based on potential business opportunities, rather than on a desire to try something new. And when appropriate investments have been made, their success or otherwise should be reviewed, and any learnings, e.g. about the investment decision-making process or the potential applicability of the technology in different parts of the organisation, should be shared.

As well as the opportunities from digital technology, the board should have a position on their appetite for digital risk. Most boards are very familiar with *risk management* around finance, cyber security and operations. Technology risk is often less well considered.

In most cases, digital technology should not be treated purely as the concern of specialist IT professionals. Ideally it will be handled flexibly, by *cross-functional teams* with flat hierarchies that allow rapid and iterative decision making. That is not to say that IT departments, where these continue to exist, don't have a role to play and won't have their own specific requirements for digital technology. But so will marketing, HR, finance and logistics. And the technology that one part of an organisation uses may well have value in another part.

The technology itself should not be reviewed, not on its own at least. Instead, *internal and external processes* that involve or that could involve digital technology, especially those that can affect customer experience and corporate reputation, should be reviewed to see whether they are as effective as they might be. Similarly, the way that competitors' products and services are delivered, and the part that technology plays in delivering them effectively, should also be reviewed where possible.

In many organisations there will be a need to improve in-house *technology skills*. The board needs to make sure that appropriate in-house skills are built up. This doesn't always have to involve employing people with those skills: other tactics such as bringing expertise in-house through seed money investments in technology start-ups can be very effective.

USING INVESTMENTS TO DELIVER INSIGHT

A privately owned British aerospace company we know of has started to make investments in digital start-ups, many of which appear unrelated to aerospace.

Much of this company's work involves the maintenance of complex and delicate machinery – aeroplanes. It employs a number of engineers and mechanics who are equipped with tools that range from simple wrenches to hand-held computers and who work in large, draughty hangars.

While much of the work is physical, the company is also a digital enterprise, with expert users of CAD/CAM software and 3D printers who design and manufacture prototype parts. The company aims to keep ahead of technological changes. And one way it achieves this is to invest in technology start-ups. It has an angel investment arm that provides small sums of long-term capital (generally £25K to £100K) for early stage deep technology start-ups that are located in the UK. The sums are small, but they spread over a wide portfolio of over 50 investments.

And the companies they invest in are not necessarily obviously con-nected with aerospace. For instance, one investment is in a company developing a robotic strawberry picker. Not much need for those in an aircraft hangar. Except that it takes sophisticated computer vision for a robot to identify a ripe strawberry of the size that is worth picking. And it takes a very precise robot for that strawberry to be picked and packed without damage. Sophisticated vision and precise handling: two tech-nologies that may have a good deal of relevance to aeroplane mainte-nance. And by investing in the strawberry picking company, they can gain insights into how those technologies work in practice and thus how they might use them in the future.

Culturally it is important that organisations are *open to innovation* and change. But more than this they should take a cross-divisional or cross-business unit view of innovative developments, encouraging ideas and knowledge to be shared between different teams. Innovation should gen-erally look to the mid-term opportunity (say 3 to 5 years out) rather than the long term so that it has some practical value to people who are likely to still be in the organisation when the innovation comes to fruition. And it is essential to allow learning, in particular learning from failure, to be valued: if a failed innovation is regarded in a negative light then any spirit of innovation is likely to die. Ensure though that there is appropriate risk management of the investments made and that there is compliance with regulations.

Strategic governance principles

Earlier in this chapter we outlined twelve strategic technology governance principles that we feel are important for organisations to adopt. We expand on them here.

Principle 1. Command of the subject

Board members should receive digital technology orientation, and they should be regularly updated on new and emerging opportunities, threats and trends. For this to happen it is likely that they will need advice and assistance from independent experts. This orientation should include an understanding of:

- How technology is used both internally but also externally (i.e. by competitors, customers and suppliers)
- Existing and emerging digital technologies that affect, or have the potential to affect, the organisation's operations
- The risks as well as the rewards that come from digital technology
- The board's, and the organisation's, responsibilities around their use of digital technology
- The integration of digital technology and the way that digital technology can have effects across organisations, beyond the area in which it principally operates.

Non-executive directors on the governing body have an important, and highly valuable, role to play here; providing other sectoral and personal experience, helping the board to avoid group think, challenging accepted beliefs and providing real support and empathy to the executives that have significant digital governance responsibilities.

Principle 2. Accountable officer

The board should ensure that a specific corporate officer (not necessarily a member of the board) is accountable for:

- Advising the board on strategic technological issues and on the organisation's ability to manage strategic digital technology issues
- Reporting on the organisation's progress in implementing the appropriate use of technology
- Ensuring that delegation processes, such as decision-making processes surrounding IT failures or ecommerce problems, are thought through (and then applied) properly
- Where appropriate, facilitating the implementation of digital technology (but, depending on their role not necessarily actioning this)
- Ensuring liaison and collaboration, facilitating joined-up actions and sharing of learnings
- Identifying deficiencies and where appropriate proposing the solutions
- Ensuring that situational awareness is maintained at board level.

Some organisations are appointing Chief Digital Officers (CDOs). Certainly, this role and title are becoming more common. If you were to decide to appoint someone to such a role, you should certainly decide where their terms of reference start and stop compared to those of the increasing number of other Chief Officers found in organisations today. And they should then be required to collaborate significantly.

Principle 3. Situational awareness

The board, enabled by the Accountable Officer, should ensure that reviews of the organisation's use of digital technology, and its readiness to use emerging digital technologies, are carried out regularly and also when potentially significant new technologies emerge:

- **Market:** The board should focus on the state of the market and in particular what competitors and technical exemplars (such as Amazon and Google) are doing; in addition, they should be aware of consumer trends and of the way that other stakeholders such as suppliers are engaging with technology.

- **Culture:** The board should particularly be aware of to what extent the organisation's culture supports an appropriate approach to the risks and opportunities provided by digital technology. In addition, they should ensure that their own influence on the organisation's culture with regard to digital technology is beneficial.
- **Compliance:** The board should always be aware of the potential for compliance issues to be raised by the adoption of digital technology within their organisation.
- **Horizon scanning:** The board should actively seek to make themselves aware of emerging technologies and be prepared to consider how these might affect the organisation's strategy in the future.

Principle 4. Ethics policies and principles

Setting and overseeing the ethical standards of an organisation is a key concern for every governing body, as it should be. After all, ethical failures can and do seriously damage corporate reputations and, in turn, harm sales, profitability and services delivery.

There are a number of ethical decisions organisations need to take when it comes to digital technology. As an example, how can you ensure that:

- You use digital technology to increase productivity without necessarily harming the well-being of your employees?
- You use digital media to increase sales without at the same time damaging the privacy or convenience of individual consumers?
- You treat the owners of the intellectual property used in your systems or products fairly?
- When machines make automated decisions about people, those decisions are fair and safe?
- Your use of technology doesn't accidentally drive ethical failures such as a lack of diversity in your organisation?
- Your use of technology promotes rather than damages sustainability within your organisation and the community, for instance by saving energy or wasting fewer raw materials?

Perhaps because digital technology is relatively new and very powerful there are still many discussions concerning what counts as ethical behaviour. Should social media companies be forced to take down terrorist propaganda online or would this be counter to free speech? Should companies (as Facebook has done) pay teenagers to download apps that will collect data about their online behaviour or is this bribing naïve people to put themselves at risk? Should AI companies create autonomous cars that promote the safety of people inside the car – and should they even try to create artificial intelligences that could one day threaten humanity?

You may not agree with all your colleagues on the answers to these, and many more, ethical questions raised by digital technology. But it is important that you debate them.

Ethics and artificial intelligence

As new digital technologies are developed new ethical quandaries inevitably arise. Nowhere is this more obvious than in the area of artificial intelligence (AI, see Chapter 14).

Artificial intelligence involves machines making decisions, many of which affect humans. Those decisions are ultimately based on choices the original programmers made. But as AI machines "learn" from their environment, the reasons for their decisions get increasingly difficult to identify. And because of that, great care is needed to ensure that their decisions remain appropriate and ethical over time.

The European Union has proposed a number of principles designed to programme ethical decision making into AI machines. These include:

- AI systems should support human rights and data privacy and should enable people to control how their data is used.
- AI systems must be able to deal with unforeseen circumstances such as errors.
- The way AI systems make decisions should be transparent.
- AI should not discriminate against groups of people.
- AI systems should be used to enhance positive social change.
- Mechanisms should be put in place to ensure responsibility and accountability for AI systems and their outcomes.

We have already seen issues with some of these principles (see the text box on Amazon recruitment in Chapter 14). How far these rules will protect humans as AI machines get even more capable of delivering complex tasks remains to be seen.

Principle 5. Information management

The information an organisation holds, explicitly in databases and documents and tacitly in the experience and knowledge of its workforce, is one of the most important sets of assets that it has. And it is one of the most difficult to manage. Consequently, information management is an important discipline. By promoting an efficient approach to identifying, capturing, evaluating, retrieving and sharing all these information assets, this discipline can contribute substantially to the organisation's effectiveness. Just imagine your organisation with all the information stripped out of it. You would probably be left with very little of value.

While the day-to-day management of information is unlikely to be high on the governing body's agenda, the process should be. And, recognising that inherent value and given the constant changes in digital technology, this process should be subject to constant review.

An effective information management system will encompass:

- **Auditing information:** What information do you have? Where does it come from? How accurate is it? What is it made up of? How complete is it? What form is it in?
- **Generating information:** What information do you need that you currently don't have? Can you use technology to collect this information and if not where else might you find it? Are there any implications to collecting or holding this information?
- **Classifying information:** How valuable is the information you hold? How sensitive is it, for instance would it damage people if it were stolen? How much is it at risk and what from?
- **Handling information:** How should you handle different forms of information? How much of it can you afford to share with others? And how can you make sure that information assets are marked in such a way that people, both inside and outside of the organisation, know how to handle them?

Principle 6. Risk management

Managing risk is one of the top concerns of the governing body and top management of any organisation. Generally, risk management is thought to encompass business risks such as strategic risk, operational risk, financial risk and compliance risk. Some people also include reputational risk here although we suggest this is a secondary risk or even an impact that results from risks and inadequate controls.

We want to see digital risk added to the list of business risks as we believe that it is worthy of consideration at the very highest level. It should not be treated as a subset of any one type of risk (such as operational risk) as it extends across all forms of business risk. Inadequate understanding of technology will obviously have a damaging effect on operations, but it can also impact on compliance (as we will see later in this book), on financial performance (not least through compliance fines but also through opening organisations up to fraud) and on the development of a strong strategic vision that will enable the organisation to compete in the future.

The board should therefore hold top management accountable for reporting a quantified and understandable (i.e. jargon free) assessment of digital technology risks and threats. This needs to be at a strategic level and relate to the organisation's objectives: a mistake many managers make is to assume that information about the day-to-day concerns they have (such as the number of cyber-attacks they have diverted) will be of use to the board.

And a mistake that many boards then make is to allow managers to give them access to all of this granular information, rather than sharing the strategic insights that can be drawn from it and upon which members of the board can then act.

Another important element of risk management is establishing what the organisation's appetite for risk taking is. Defining this is ultimately a responsibility of the board. Unfortunately, the risks surrounding digital technology can be difficult to define and build a consensus upon.

- A data breach involving personal data could involve a hefty fine from a European regulator. Or it could be handled by the regulator offering some advice. This might depend on the scale of the breach which would be impossible to predict. (After all, if you could predict it, then it probably wouldn't happen.)

- The risks from implementing an emerging technology may simply not be understood.

How do you define risk appetite under those circumstances? The reality is that for digital technology it will probably be difficult to do so with any degree of precision. But a series of plain language statements can be made that, for instance, might indicate a different appetite for the website being attacked by hackers and for the customer database being attacked. These statements, together with an analysis of identified risks, will help you understand whether any risks are outside the parameters your organisation wishes to accept. When such a risk is discovered, action to mitigate the risk will of course be needed but relatively straightforward to do, e.g. "We will, from now on, not allow...."

Finally, it's important to accept that digital technology risks are not simply the same as cyber security risks. Yes, cyber security is important but digital technology comes with many other risks that are not directly related to information leaks. We explore this issue more in future chapters but particularly Chapters 11 and 12.

Principle 7. Appropriate technology use

Digital technology brings with it many opportunities and the board should ensure that appropriate investments are made in support of the organisation's purpose. This will include ensuring that:

- The board, and the organisation's top management reporting to it, actively demonstrate leadership in, and commitment to, the use of appropriate digital technology consistent with the organisation's purpose.
- There is board-level acceptance of the role of digital technology plays, what risks it brings to the organisation and how far there is an appetite for further digital transformation.
- The board takes an active role in the planning of digital technology and ensuring that it is a prime consideration within the organisation's strategic annual planning.
- The necessary resources are made available to support the efficient use of digital technology.

- The integration of digital technology into wider systems is addressed, e.g. the process of communicating information from a website to a worker picking and packing goods in a warehouse, or from a High Street shop to customer care call centre.

Principle 8. End-to-end customer experience

One major benefit of digital technology is the way that it enables organisations to engage with consumers successfully, and to keep an accurate record of those communications that can be analysed. Compare this, for instance, to a corner grocery shop keeper who can only talk to customers when they enter the shop and (probably at least) has no way of recording those conversations.

THE IMPORTANCE OF CHOICE IN CUSTOMER EXPERIENCE

Ensuring an excellent customer experience is fundamental to delivering successful products and services. This isn't a difficult thing to analyse, although it does take a little effort. Often it simply involves a common sense appraisal of what typical users will experience. However, not all users are the same and an essential element of an excellent customer experience is choice.

Here is a simple example of how services can be let down by a failure to think through the end-to-end customer experience. Recently, when planning a visit to London, I made an online booking for my car so I could park it at 8.30 in the morning in a commercial car park, prior to getting public transport into the centre. I wasn't sure when I would return that day, so I purchased a 24-hour ticket. Later that day I picked up my car and drove home. I was, however, irritated to be woken at 4.30 the following morning, by an automated telephone call, telling me I had 4 hours left before I would need to book additional parking. Why was it necessary to call me? Certainly, the option of a call might have been useful to some people; but to assume that I needed one (especially one at 4.30 am) showed considerable short-sightedness. It would have been nice to have been given the choice.

Wherever possible choice should be built into the customer experience. An important area is in email marketing. With consent (see Chapter 12)

> often being used to justify emailing commercial messages to customers, the frequency of those messages is an area where choice can be offered. Some people may want to hear from you every month, others every week or even every day. If you want to keep your customers happy, giving them the choice of when they hear from you is a good tactic.

This benefit brings with it an enormous opportunity: the chance to develop an end-to-end plan for delivering excellence in customer experience – before the customer buys, during the buying process and subsequent to the sale being made. Not only that, customers can potentially be tracked across different purchases and can be up-sold and cross-sold as they buy, with communications and offers based on an analysis of previous behaviour and the behaviour of similar customers. It is the responsibility of the board to ensure that this opportunity is thoroughly investigated.

However, if this opportunity is to be grasped then the practicalities of the user experience must be addressed through the formal and regular analysis of any user interfaces designed for consumers (known as user testing or usability testing). This is because an excellent user experience, sometimes evidenced online in a simple and intuitive interface (that will have taken rather more effort to achieve), is a crucial underlying success factor for digital products. It wasn't a coincidence that Google overtook Yahoo as the most popular search engine. Nor was it a coincidence that Facebook toppled MySpace as the leading social media platform. In both cases, a first-rate user experience did the damage to the incumbent.

Improvements to the use of digital technology to provide excellence in customer service should be implemented on a continuous basis. We explore these issues further in Chapters 4 to 6.

Principle 9. Community engagement

Different people have different experiences of, and different knowledge of, digital technology. For that reason, the board should encourage collaboration with other organisational stakeholders, especially employees, customers and suppliers but also including:

- Experts such as academics and journalists
- Standards bodies
- The technology working parties of industry bodies
- Technology groups including those set up by government such as the UK's Digital Catapult.

In order to ensure that digital technology is best used in support of the organisation's purpose, the more different viewpoints the board can encourage into the organisation, the more successful their organisation is likely to be.

Principle 10. Resilience planning

Organisations are increasingly dependent on digital technology. Twenty-five years ago, if an organisation suffered the theft of all their computers' memory chips (as did happen to an advertising agency we know of) they could get by with faxes, typewriters, telephones and face-to-face meetings. It would be different today. And that's why resilience planning is so important.

To enable this to happen, the board should ensure that top management supports the creation, implementation, testing and ongoing improvement of plans for the use of digital technology in support of the organisation's purpose and importantly plans for eventualities where that technology is compromised in some way, such as:

- A major data breach or an attack on a critical website
- A power outage
- The failure of a cloud computing supplier
- Malicious software that prevents certain IT systems or computer-controlled machines from working effectively.

We explore resilience further in Chapter 13.

Principle 11. Effectiveness and continuous improvement

The board should periodically review its own performance in ensuring the appropriate implementation of digital technology across the organisation through:

- **Monitoring and reporting** of the degree to which employees (and fellow members of the governing body) follow policies or rules relevant to digital technology, such as cyber security or IP licensing rules; and, if not, then why not.
- **Learning lessons** about the way that digital technology opportunities are identified and taken up, the way that any associated risks are identified and mitigated, and the way any potentially damaging incidents are managed. As part of this, an awareness of how competitors and other businesses are engaging with technology will be helpful.
- **Continuous improvements** in the way that top management are seen to integrate digital technology into organisational operations and their motivation for delivering continuous improvement.

Principle 12. Flexibility

The final strategic principle is that of flexibility. Leaders need to make sure that their organisations remain agile with regard to digital technology, are able to adapt to new technology, are prepared to abandon old technology when appropriate, despite any sunk costs in capital assets that have yet to be fully depreciated away, and are emotionally capable of abandoning experiments with technology that appear to be failing.

To achieve agility, they will need to be able to question their own strongly held beliefs and to encourage others to do so.

Perhaps most importantly, they will need to encourage a culture of innovation, whether that is through old-fashioned suggestion boxes or shiny new online innovation platforms. Because whatever technology is used, it is the culture behind innovation (especially at middle management levels where many obstacles can arise) that will dictate whether innovative ideas thrive in an organisation or are strangled at birth.

Delivering strategic change

Twelve different principles are a lot to take on. But we believe they are all important. If organisations adhere to each one, to the extent that they can or wish to, then they can be confident that they will be operating within a structure that promotes the degree of agility and quality that they have identified as essential for the prosperity sought and that takes advantage of the many opportunities that digital technology offers. It is only through this sort of discipline that organisations will be able to create and achieve the vision of digital transformation that so many, rightly, strive for.

One of the major issues for delivering strategic transformation is the need to manage operational change within the organisation and we address this issue in the next chapter.

3

MANAGING RAPID CHANGE IN A DIGITAL WORLD

Summary

Technology moves very rapidly with power and functionality increasing as prices plunge. While this is undoubtedly an opportunity it is also a problem: why invest now if prices will be better in a few months' time. This is then a recipe for stagnation and ultimately market irrelevance. Organisations need to accept that constant technological change, via evolution rather than revolution, is the answer.

Constant change is difficult to manage. It requires a deep understanding of the current situation as well as insights into likely futures. And working towards those futures needs careful planning combined with the flexibility to change direction if necessary. Success will come mainly from three things: understanding people (employees and customers); being innovative; and reaching for excellence. With that philosophy in place, individual projects, or programmes of linked projects, can be implemented. But their success will only come if governance controls are combined with the freedom to think differently and to fail well.

Digital technology moves at a blistering pace and it's hard for organisations to keep up. Boards need to ensure that their organisations adopt proven and effective technologies while avoiding attractive fads that in reality are little more than technologies looking for a problem to solve.

But identifying the best technologies to adopt isn't the only problem. Implementing them, even as those technologies continue to evolve, is difficult. It's easy to get distracted as new opportunities appear from different technologies. The temptation may be to abandon one implementation in order to grab another, which in turn might appear to be outdated before implementation is complete. It is important to avoid this because the benefits of adopting digital technology are very real, potentially worth trillions to the global economy.

The digital opportunity

The potential benefits of well-planned and managed digital technologies are likely to be substantial. The World Economic Forum's digital transformation initiative has estimated that $100 trillion of value could be created by digitisation in global manufacturing over the next decade. And the UK Government's *Made Smarter* review (1) estimates that a "positive impact of faster innovation and adoption of industrial digital technologies could be as much as £455 billion for UK manufacturing over the next decade".

Made smarter outlines a number of technology-enablers, together with the associated benefits that these technologies will bring to manufacturing (Table 3.1).

The exciting thing is that as well as manufacturing, all industry sectors, including service industries and the public sector services, will be very positively impacted by the same digital enablers.

In this chapter we will look at how organisations can cope with managing technology in a rapidly changing world.

The pace of change

On 20 July 1969, two brave men, Neil Armstrong and Buzz Aldrin, took humanity's first steps on the moon. And while they were undoubtedly heroes there was another hero we don't hear about so much – the Apollo Guidance Computer. This wonderful piece of technology took humans to

Table 3.1

Technology enablers	Benefits
Advanced robotics	Flexibility
Additive manufacturing	Productivity
Augmented reality	Speed to market
Simulation (and digital twins)	Quality
Industrial internet (and sensors)	Scalability
Cloud computing	Innovation capability
Big Data and analytics	Robustness
	Cost reduction
	Sustainability
	Safety
	Better working conditions
	Continuous learning
	Collaboration

the moon. And it was about as powerful as the simplest of today's pocket calculators.

The last 30 years has seen enormous change. In 1993, the arrival of the worldwide web, and in particular the ability to display pictures on web pages, started to get people (and organisations) interested in using the internet. It turned it from a place where a few hundred thousand people discussed cats and baseball in clunky discussion groups into a place where millions of people went every day to shop, talk and be entertained.

The arrival of wireless internet access changed the way people use the internet further, freeing people from terminals attached to phone lines in homes and offices. Smartphones completed the change, making internet use even more portable and convenient.

Now well over half of the world's population (and in Europe and North America that figure is closer to 90%) (2), watch TV, play games, shop for groceries and post pictures of what they had for lunch.

During that time, the price of this technology (which would have seemed almost magical 30 years ago) has plummeted. A standard personal computer in 1990 cost around $2,000. Today you can get fast internet access on a smartphone for under $20. And at the same time, of course, the power of these devices, and the storage space available on them, has exploded.

This has had, obviously, a massive effect on businesses. International communication, information storage, data collection and analysis, and many business processes have been transformed as we will discuss in subsequent chapters of this book. But equally importantly, for business, society has been transformed. People who have grown up with technology take it for granted and demand it (including what it can deliver – entertainment, communication, information, money) wherever they go. These people, aged 30 or under, make up over half the world's population and, even in Europe account for a third.

And older people are going online too, in ever-increasing numbers. In the UK, the proportion of internet users in the 65 to 74 age group increased from 52% in 2011 to 80% in 2018 – not quite the 99% seen in adults under 35, but still a massive change.

Businesses, at least some of them, embrace this transformation in their markets, while governments struggle to keep up.

The problem with technology

It sounds like a wonderful opportunity. Businesses can deliver the products and services that people have always wanted (even if they didn't always realise it) – customised, instant, convenient and available anywhere.

But that doesn't mean it's an easy opportunity to manage. Being at the bleeding edge of technology is expensive, risky and unlikely to pay off in the short term. Tesla, Uber and Spotify all made losses in 2018. Twitter only struggled into profit after 10 years. And even Amazon, profitable in its home market, reportedly loses money on its international business.

But perhaps these problems are limited to technology companies as opposed to companies that use technology. And it does appear that for companies operating in the real world of shareholders, tax and regulations, technology investments generally do pay off. Indeed, some estimates (3) indicate that 80% of companies that engage with digital transformation increase profits.

Unfortunately, other people (4) estimate that in over 80% of cases, digital transformation initiatives fail. Why might that be?

The truth is that managing technology is hard. And managing rapid change is hard. Add the two together and you are faced with a considerable challenge.

Managing technology

Managing technology is difficult, especially for people who are not technologists. Sometimes it seems that technologists speak a totally different

language full of TLAs (Three Letter Acronyms). They have different atti-
tudes to risk, deliverables and even time from their non-technical coun-
terparts. They think in different ways, far more literally than most people.

Well, that may or may not be true. But the difficulty in managing technology doesn't stem from the nature of the people involved. If you can't
manage them, that's more likely to be a reflection on you, not on them. It's
to do with the nature of technology, which is:

- A catalyst for new ways of doing things
- Constantly changing
- One step removed from end products.

Technology (especially digital technology) is often a catalyst for new ways
of thinking and working. And there may be no one with experience of these
new ways in your organisation, or if there is, perhaps you do not trust them
to match the promise of the technology with the needs of your organisation.

Because technology constantly and rapidly changes, in ways that are
sometimes hard to foresee other than generally getting better, smaller and
cheaper, it's hard to know whether the decisions you make today will still
be appropriate tomorrow.

And because technology generally isn't the product, but instead powers
the product, the aims of technology won't always be the same as the aims
of the product, and success will be measured in different ways.

Managing rapid change

Managing change is always hard and managing rapid change is particularly
difficult. Decisions, which may need considerable thought, need to be made
quickly and circumstances may have changed before the decision is made.
And if circumstances (including technology) change so rapidly, the temptation is to avoid making any decisions in the hope that the pace of change
might eventually slow, allowing reasoned decisions to be made. That is a recipe for suicide – as any frog who is slowly being boiled alive will tell you (5).

Change is difficult for many reasons:

- It is about moving forward while much of governance is about using
  backwards-looking tools such as financial accounts. But just because

something worked for you last year that doesn't mean it will work well for you this year.

- Organisations resist change. Most people are cautious, and most people like to remain comfortable doing and using what they know. But organisational change has to be started before it is too late. And in a time of rapid change that's a problem as the technology required to deal with changing circumstances may not yet be mature.

- People don't just fear change: they fear technological change. They think "I am going to lose my job to a robot" or "I'm going to look stupid because I don't understand this new stuff". And because they are frightened, they look for excuses not to change: "I'd never use that, so I don't think anyone else would either".

Getting started

It's inevitable that technology will change and bring changes to society along with it. You might as well accept that and start making decisions about how you are going to react to technology, which will either be by changing your products and services or by changing the way your organisation delivers those products and services.

Watching out for the risks

But before you make any decisions, it's a good idea to remember that introducing new digital technology carries many risks that may be difficult to manage. Why is that? There are a number of reasons:

- The risks may not be well understood if they haven't been experienced by anyone on the team. And with any new technology that could well be the case.

- The skills required to manage those risks may not be available in-house. You may be dependent on long-term relationships with third parties and contractors. Inevitably they will not understand your organisation as well as you do, although they may well think that they do. And they will be more loyal to themselves than to you.

- If you are using new technology, your project may be experimental with fuzzy goals and outputs that have not been defined precisely because they are hard to visualise or anticipate.

- People involved with the implementation of the new technology may make assumptions that you do not hold, or they may use jargon and words that mean different things to different people. Misunderstandings and disappointments can easily arise, eroding trust and damaging team spirit.
- The normal iterative change processes that many organisations used to use to implement change can be far too slow to be effective when technology changes so fast.

Conversely, another common experience is that the proposed change is implemented successfully but lacks ambition, perhaps because of fear or of a lack of imagination about what might be achieved. This isn't, of course, a problem experienced by most start-ups. It's easy to visualise a new machine. It's far harder to visualise a new end-to-end process. There is a danger that the focus of digital transformation can be on making a physical change, in effect bolting a computer onto a system to make it a bit more efficient, rather than thinking how the system as a whole could be made radically more efficient and effective through technology.

This problem is often made worse by a focus on processes or even a particular endpoint, such as a product or output from a process, rather than a wider focus on continued business success.

And just to make things a little worse, the implementation of technology is often driven at speed, perhaps because it was started late and people are panicking, perhaps because the enthusiastic developers want to ignore due process (and/or compliance) and get to a result as soon as possible.

None of this makes implementing technology easy.

Managing technology risks successfully

Let's start by thinking about what you shouldn't do when you are transforming your organisation digitally.

- Don't set up a digital transformation office or hire a digital chief and leave them disconnected from the rest of the organisation: that way they will make a lot of noise but achieve very little of practical value.
- Don't try to digitise all your processes without thinking about where your priorities should lie and where the most value is to be

found. If you do that then you may spend a lot of effort in changing something that doesn't really need changing or add much when it is changed.

- Don't focus on technology at the expense of the end users (whether they are employees or customers) because if you do your end users will simply reject your technology.
- Don't think that a single change like developing a responsive website or digitising your call centre is the same as transformation because if you do you will have fooled yourself into inactivity. See also Chapters 4 and 5.
- Don't be over-ambitious: if you try to do everything at the same time you will probably fail at everything.
- Don't reorganise and then, when you worry that things are not perfect, reorganise again returning to where you were 5 years ago (we are thinking of certain major UK national organisations here).
- Don't think that digital transformation is easy.

Well, that's what you *shouldn't* do. But there are plenty of things that you *should* do to lead your organisation towards digital transformation.

Understand your business environment

The organisation's governing body needs to have a firm understanding of the realities of the business environment they operate in. And of course, you do have that understanding. But can you honestly say that your understanding extends as far as the implications of digital technology?

Take a look at the inside of your organisation, the technology assets it has and their relative value to you. Do you have market data that isn't being fully exploited? Could some of your customers be served in different ways? Are there digital skills within the organisation that are under-used?

Take a look externally too. Use technology such as data analytics to understand your target markets but don't ignore your own personal experience or the opinions of the people around you, at home and at work. Treat yourself and those around you as a sort of mini focus group: try to understand why people do things the way they do when they interact with technology. When you do this, however, you should remember that your life may be different from the lives of other people (not everyone lives in a city with fast data connections and – odd though it may sound – not everyone has an iPhone).

At the same time recognise the changes that digital technology can bring to your operational environment through:

- **Automation** of processes. With digital technology many processes can be completed more rapidly and with higher quality. Humans can potentially be freed up for more creative work that adds more value. Is your organisation making the most of your human talent or tying them down in routine procedures that could better be done by a machine?
- **Connectivity**. People, and machines, are increasingly connected. How can you take advantage of the fact that your customers can connect to you when they are travelling and at night as well as when they are at home or at work? How can you use that fact to offer a better service or an additional service? And how can you ensure that your employees are taking advantage of the benefits of mobility to work more efficiently and with greater job satisfaction?
- **Data**. There is more and more data. Much of it is valuable but some of it is just noise; and some of it is actively misleading. How can you tell these apart? And when you have uncovered the most valuable sets of data how can you combine data to get greater insight into the solutions your organisation seeks?
- **Speed**. The processing and communication of data are getting faster and faster. Near instantaneous data transfer is crucial if you work in currency trading. But if you don't it is still a vital advantage. Employees and customers who get near real-time updates to the information they are looking at on their computers will get less frustrated and will be more likely to behave in ways you want them to. What advantages does fast data transmission bring to your organisation?

Remember though that there are undoubtedly many practical constraints on your use of technology. Some may be real, but others may be perceived. You may not have the resources or knowledge to take advantage of artificial intelligence. But perhaps you do have the ability to make better use of data analysis.

And remember also that your consumers and employees are only human. They need technology that works for them. Some technology that sounds enticing isn't market ready; some is badly designed; some may still be too expensive or relatively untested.

Provide active leadership

People who are part of the leadership of an organisation should be proactive when it comes to the implementation of technology, not reactive. They should seek out reasons for change and promote them, rather than waiting for a crisis to hit.

At all levels of an organisation, managers may feel they are too busy with day-to-day concerns to focus on what might happen in the future. This is dangerous. Firefighting is all very well, but if there is a tsunami of technology around the corner then any amount of firefighting is simply wasted effort.

Of course, not all digitisation projects will require your direct involvement. Some will, especially when they involve significant investment, major changes, major risks or relate to the fundamental objectives and mission of the organisation. But even where they don't it is still your responsibility to ensure that the organisation looks to the future, rather than simply to the recent.

Organisational leaders can be particularly effective at influencing two factors that will bear on the way an organisation engages with technology: culture and hierarchy.

- **Culture**. The leaders of organisations have a huge influence on culture. While not all culture comes from the top, a good deal of it does. Certainly, the importance that the organisation attaches to innovative technology will largely come from the stated opinions and behaviours of members of the board.
- **Hierarchy**. If the innovative use of technology is to thrive, individuals within an organisation need to be given the freedom to act. It is a paradox of governance that in a time of change, flatter organisational structures are needed to provide this freedom but at the same time, strong governance is also needed to retain control.

While engaged in providing leadership you will probably need support. And this is where having experts on the board can be a real benefit.

Non-Executive Directors with experience of how competitors, or even businesses in entirely different industries, have used digital technology will be invaluable.

Getting digitisation right

Have the right objectives

Leaders need to understand the business value that would be brought by digitisation and focus on measurable objectives. It's all too easy for digitisation to focus on the technology rather than on the benefits to the end user. Don't be swayed too far by technology enthusiasts: always focus on the value to the business and be on the lookout for anything that indicates that value may not be delivered.

This means prioritising the opportunities. It's likely that innovative proposals to use digital technology will come from various parts of the organisation. You can't do everything, not at the same time anyway. So there will be a need to agree where value comes from and where it is perhaps eroded. Thinking about your organisation's biggest weaknesses and pain points and where the biggest threats are located helps you to prioritise.

And once you have prioritised the opportunities, stay focussed on them. Don't get distracted every time a new opportunity – or, even worse, some sexy new technology idea – comes along, or you will end up doing nothing.

Plan for change

Planning to implement a new digital technology is no different from planning for any other major organisational change. The degree to which you, as a leader, get involved will naturally enough depend on the significance of the opportunity and the nature of your organisation. You should:

1. Identify and define the issue and ensure the whole leadership team accepts it exists and needs solving urgently.
2. Prioritise the changes that are needed and limit your ambitions largely to the most important; at the same time accept that some early, easy, but perhaps trivial wins may still be important to add credibility to the process.

3. Map the stakeholders involved in the change; identify the potential champions and resistors; ensure that the champions are given sufficient influence to aid the change while the resistors' perceptions and fears are addressed through methods such as education, changed incentives or even ideally involvement in the change process.

4. Map your organisation; identify potential obstacles to success (such as silos, a lack of resource or conflicting incentives) and ensure these are addressed as far and as fast as possible.

5. Focus on the first 100 days by when you should be able to prove success is on the way, and after which you can take a more hands-off approach.

Promote the right structure

For any major changes, you will need to ensure that the right organisational structure is in place.

- **Obstacles:** Look for structural obstacles that might impede the use of technology. Is your organisation's data held in different silos that make analysis difficult? Is the IT infrastructure inadequate to deliver the speed of data transfer that your employees need to work efficiently? Do you have conflicting incentives in place that actively encourage employees to sabotage digital change?

DISINCENTIVISING DIFFERENT TEAMS

Around the turn of the millennium I worked for the digital arm of a publishing company, developing online advertising revenues. I met an advertising client who I knew had recently been talking to one of the press sales team at a conference. "I didn't realise you took online advertising" the client said. "I was talking to George last week and he never mentioned it". Mildly irritated, I took this up with George next time I saw him. "You might tell the clients about our new websites", I told him. "Never", he replied. "If I can do anything to stop you making a sale I will". I was a bit taken aback until George followed up with: "Each time you make a sale that's less money a client has to buy from me. And that's commission

> straight out of my pocket". Quite understandable really, but it was unfortunate that the way that the organisation had implemented digital technology meant that George and I, while we worked for the same company, were now direct competitors.

- **Resources:** Ensure that the resources needed for success are there. Does the organisation have the skills available, internally or externally, to achieve success? Are the people involved in the trialling or implementation of digital technology given sufficient bandwidth to make things happen and are their managers content with this? And is organisation's commitment to digital technology sufficiently robust to continue with it despite any short-term revenue dips?
- **Integration:** It's important to remember that new digital processes may need to be integrated with non-digital processes or data and with legacy digital processes that may store data in different formats or use different software. Without proper attention to integration, users may find that they have to repeatedly enter the same data in different parts of a process. Anyone who has laboriously input their account details into an Interactive Voice Recognition (IVR) system "so that we can provide you with the right help" only to be asked for the same data by a human operator will understand how irritating this is. Equally, developing an application that allows any input into a field, rather than only the permitted formats (so as to prevent processing rejections later), will upset members of your teams who don't even have a direct influence on the new system that provides their inputs.

Avoid short-termism

Take a mid-term view or even a long-term view of customer needs, rather than a short-term view. It may take time and experimentation before the anticipated value from the technology is delivered. Digital technology changes constantly and customer needs and wants change with it. A programme of digitisation aimed at responding to immediate problems is unlikely to be successful.

Avoid unnecessary risk. For most organisations it is better to avoid being a pioneer. First mover advantage is rarely an advantage online and it is best to let others make the basic mistakes before you pilot your own ideas. And do pilot them. Going for incremental development and launch is normally better than looking to make a big bang.

First mover disadvantage?

Many people feel that the first organisation to develop a new service or product has an enormous advantage over others. But does first mover advantage really exist? Take Henry Ford as an example. He didn't invent the motor car and he wasn't the first person to use a production line. But he built up a very successful enterprise nonetheless.

Does first mover advantage exist in the digital world? Perhaps, but it's by no means universal (Table 3.2).

Table 3.2

Product	Leader	Date	First mover	Date founded
Search engines advertising	Google	2002	Overture	2000
Online shopping	Amazon	1995	Netmarket	1994
Social media	Facebook	2004	Six Degrees	1997
Web browser	Chrome	2008	Lynx	1992
Word processing	Word	1984	Wordstar	1978

Take account of people

You need to assess the likely effect of new technology on the employees and the customers who will use it or be affected by it. If customers experience a lower level of service, for instance when a human operator is replaced by an IVR system, the negative effect of that needs to be balanced against the positive savings delivered.

But, if there are negative effects then ways of managing them should be sought, as they are not themselves immediate justification for not making the change – they are just part of making it. For instance, the inclusion of artificial intelligence in an IVR system can enable the system to detect (from words used or tone of voice) when a customer is becoming frustrated. In that case, not only can the customer be immediately moved to a human operative, but the existence of frustration can be used to improve the IVR for future users.

Organisational culture will have a large effect on whether digitisation is successful. Managers and workers may be unhappy with change to established processes that they may feel comfortable with. Don't assume that everyone under the age of 30 is eager to try new technology. Some will be (as will some of your older workers) but some won't. And if there is a risk that employees find their job is less rewarding (because they have less human interaction), or frustrating (because the technology doesn't quite deliver what you wanted it to), you will need to take account of that too. A good tactic to employ would be to involve real users in application and system testing. And even when the application development is complete, be sure to at least consider whether it would be possible and beneficial to undertake a piloting exercise then deployment to a pioneer group and then a staged roll-out. This will then engage your users, allow them to show up issues (and then help you fix them) and therefore feel valued and fully engaged in (ideally, even absolutely critical to) this successful change.

Encourage innovation

Promoting a culture of innovation is essential if you want your organisation to make the most of digital technology. It's impossible for one person to be aware of all the possibilities that digital technology can bring. Many of the best opportunities will be identified by people who are performing roles that they find frustrating because they know new technology could help them do a better and more efficient job.

Given its importance in a time of rapid change, it's unfortunate that many ordinary workers do not feel free to innovate. In fact, in the UK, although the vast majority (91%) of employees say they want to innovate at work, only 34% feel free to do so (6).

Innovation presents a problem for governance. Intrinsically, it means the freedom to try new ways of doing things. How does governance, concerned as it is with defining policies, processes, efficiencies and results, deal with a business process that involves the unknown? Innovation requires that people are given freedom; governance requires that they are not.

That's an entirely false paradox of course. But it does illustrate a point of difficulty perhaps. Leaders need to define the way their organisations treat innovation; they need to put boundaries around the amount of freedom people are given and to ensure that the processes that support innovation are robust, including having processes for killing off projects for new products and processes that are failing. Like much in life, it's a balancing act. But it's one that can and must be managed, if the board is prepared to ask questions such as:

- Do we have an innovation strategy, and do we discuss it regularly?
- Do we review the progress of the most important new ideas within the organisation?
- Have we considered and agreed on our appetite for innovation risk?
- What goals around doing things differently have we set for top management, and how do we measure and review these?
- Are we getting regular reports on how we are developing new ideas in the organisation, and are we taking these seriously?
- Do we encourage people to think differently and support them when they come up with new ways of achieving our corporate goals?

That last question is perhaps the most important. Innovation can be encouraged in a number of ways:

- **Horizon scanning:** Ensure that your organisation is monitoring changes to technology that might be significant for your organisation. Any important changes, and their potential significance, should be reported to, and explained to, the board on a regular basis (and that doesn't mean once a year).
- **Engagement with innovators:** Innovators are everywhere. And while some will be in your organisation most won't. Engaging with innovative thinkers, journalists, academics and entrepreneurs will bring new ideas into your organisation.
- **Innovation systems:** Formal innovation systems can be implemented. These can work in a number of ways. Internal teams can be specifically tasked to come up with new ideas or solve existing problems. External specialists may be used, for instance, competitions can be run where problems are stated and prizes given for the best solutions provided by members of the public. (This is similar to the bug bounties that many large software companies provide to people who uncover faults or vulnerabilities in software.) And digital innovation platforms can be used. These are systems that combine electronic suggestion boxes with

project management software and they can be used to collect ideas, filter out the best ones and assist with their efficient development and implementation.

- **Inclusion:** Flatter hierarchies can be developed where people feel included and valued. Many would claim flat hierarchies or even "holarchies", where individual workers or teams have their own roles but cannot on their own influence other workers or teams, are essential for innovation as people need the freedom to act on their ideas. This isn't the same as allowing a complete free for all. You will still need to ensure that the appropriate processes for implementing new ideas are in place. But the right sort of structure, where teams have the flexibility to act and to organise themselves, can be very beneficial.

- **Training** that keeps the skills of workers up to date. If people have access to the latest information and thinking they are likely to have the knowledge to come up with creative and innovative ideas.

- **Change champions** can be used to spread the word. It doesn't matter how senior or junior these people are, so long as they have presence and the associated influence. The newest intern fresh from university may well be a highly respected technology guru even if they are unable to write a coherent presentation or get to work on time. In some cases, a change champion can be supported by a centre for digital excellence where people can go to explore the latest technology and exchange ideas about how it might be used.

- **Incentives** can be used to reward people who come up with new ideas and different ways of doing things even if those ideas don't always work. It's important to reward innovative thinking itself and regard failure not as a negative result but as a way of learning about what would work better.

- **Flexible working:** Allowing people to work remotely when they need to and encouraging a healthy work-life balance promotes new ways of thinking, at least because people are more likely to be supportive of their organisation if they are treated well.

Finding useful innovation will always be challenging. And it's important to emphasise that innovation isn't just about having lots of new ideas: you need ideas that have a purpose, a business application, that have potential to make your organisation more effective. Anyone can have ideas. Having commercially effective ideas is the hard bit.

Reach for excellence

Do it well. Simply using technology isn't sufficient. It has to be used well. And that means:

- **Careful design**, oriented towards users, not developers. You will need business rules and decision trees, clear agreement on what functionality is needed, robust end user ("usability") testing, identify where you pilot the new technology and the people that you can persuade to be pioneer users of the technology when it is finally rolled out.
- **Holistic design** where you think about the whole of the customer experience. Amazon's vision is to be "Earth's most customer-centric company" (7) and you can argue that they deliver that from their easy to use website, their one-click buying (great for me, even better for them), their massive range of products, their flexible delivery options and their matchless returns policy.
- **Compliant design** where important elements such as security by design and privacy by design are an inherent part of the development process and not bolted on as an afterthought.
- **Inclusive design** where everyone can access the service. As part of that, ensure you design your services for the digitally dispossessed, those millions of people who don't regularly use the internet or don't even have an email address or even a PC (and that's over 5 million in the UK, a sizable number).

Monitor success

Engaging with some or all of these ways of promoting digital technology is admirable and we'd like to think after finishing this book you do start putting some of these things in place.

But to succeed you will need stamina. A one-off fit of enthusiasm won't get you very far. You need to measure progress, you will need to continue to communicate the benefits that *are* being achieved, and you will need to keep going through the Slough of Despond when things don't work out, as sometimes will happen.

Projects or programmes?

Organisations that decide to embrace digitisation often initiate a large, rather monolithic, programme of change, headed up by a digitisation team under a Chief Digital Officer (CDO) or similar and with an administrative burden of recording and reporting from overly elaborate project and programme tracking. This can be a mistake as these large programmes may be unwieldy and hard to manage successfully. You just have to consider some of the UK government's IT initiatives (the NHS's abandoned patients record system that cost the tax payer some £10 billion and the painful deployment of Universal Credit come to mind) (8).

Worse, monolithic programmes can crowd out smaller innovations elsewhere in the organisation which then get seen as being off strategy as they are not part of the agreed big picture even when they may make a major impact upon a critical process. And then, purely because they are seen as distracting, they are killed off or given unreasonable success criteria, which strangles them after a few steps.

Instead, you need to be more flexible. Formal programmes can work if they are a collection of smaller projects that are mapped together with the intention of delivering certain outcomes. Smaller projects, while individually less ambitious, are generally lower risk and more likely to succeed because they are easier to specify, design, build and manage. They are also more easily absorbed into relevant programmes and can add real value/ benefit to that executed programme.

But, if one project within a programme is failing, this also won't then hold up *all* the other digital projects in the organisation and will only have a small impact on the programme within which it sits.

Alternatively, an *ad hoc* approach can be taken with innovative projects welcomed from across the organisation, whatever technology and whatever the end goal.

By this, we are *not* arguing for anarchy here. Digital projects do need some control and it will generally be sensible to appoint someone with experience to oversee the integration of digital technologies and the various digital projects into your organisation. This person should not be responsible for deciding what digital technology is used within the organisation. Instead, they will have a role that involves:

- Acting as an information resource for the people who are leading individual digital projects, for instance helping them to build businesses cases
- Identifying (and helping to manage) project, programme and strategic risks
- Supporting the culture of innovation and technology engagement across the organisation and acting as a champion for technology change
- Identifying priorities for the organisation, especially when resources are limited and when projects are found to be clashing
- Putting in place consistent processes for tracking delivered value so that the success or otherwise of individual projects can be compared with others and the overall benefit derived from digital projects identified and reported
- Facilitating understanding, prioritisation and mitigations when a technology project that is bringing benefits to one area of the organisation also brings problems to another
- Facilitating the integration of successfully implemented and therefore proven new technology from one area of the organisation into other parts of the organisation
- Enabling the organisation to "fail fast" by acting as the executioner of projects that have obviously failed
- Supporting the teams responsible for failed projects by defusing any blame that may be attached and emphasising to the wider organisation the learning that has come from it
- Sharing learnings from successful projects across the organisation and seeking learnings from outside it
- Being the prime external communicator on behalf of the organisation of its digital change successes.

This is a challenging role to take on, but an important one. For it to succeed it will definitely need the full support of the board.

Managing change is an essential business skill and you will need to ensure that the top management in your organisation are ensuring that change is managed appropriately, enabling your organisation to remain flexible. And one of the most important areas to assure change is in the internal processes that power your organisation. We consider how these internal processes can make the most of digital technology in the next chapter.

References

1. Maier, J. (2017). *Made Smarter Review*. London: Department for Business, Energy and Industrial Strategy. [online] Available at: www.gov.uk/gove rnment/publications/made-smarter-review [Accessed 24 May 2019].

2. MiniWatts Marketing Group. (2019). *World Internet Users Statistics and 2019 World Population Stats*. [online] Available at: www.internetworldstats. com/stats.htm [Accessed 24 May 2019].

3. Westerman, G. *et al.* (2014). *Leading Digital: Turning Technology into Business Transformation*. Boston: HBR Press.

4. Michael Gale of PulsePoint Group, interviewed by Bruce Rogers. Rogers, B. (2016) *Why 84% of Companies Fail at Digital Transformation*. [online] Available at: www.forbes.com/sites/brucerogers/2016/01/07/why-84-of-c ompanies-fail-at-digital-transformation/#7ac5b304397b [Accessed 24 May 2019].

5. A vivid image popularised by Charles Handy in his 1989 book *The Age of Unreason*, where he described how a frog in a kettle might allow itself to be boiled to death if the water (a metaphor for change) didn't heat up too quickly.

6. Consultancy.org. (2019). *Two Thirds of UK Employees Not Empowered Enough to Innovate*. [online] Available at: www.consultancy.uk/news/2 0747/two-thirds-of-uk-employees-not-empowered-enough-to-innovate [Accessed 24 May 2019].

7. Amazon. (2019). *Amazon's Global Career Site*. [online] Available at: www. amazon.jobs/en/working/working-amazon/#our-dna [Accessed 24 May 2019].

8. Syal, R. (2019). *Abandoned NHS IT System Has Cost £10bn So Far*. [online] The Guardian. Available at: www.theguardian.com/society/2013/sep/18/ nhs-records-system-10bn [Accessed 24 May 2019].

4

DIGITISING INTERNAL OPERATIONS

Summary

Most organisations have begun to digitise at least some operations. But there is often huge potential for further digitisation right across organisations, in finance, HR, sales, marketing and operations as well as IT. Often this will be driven by a desire for efficiency and to reduce waste. But it can also be driven by a desire to improve quality and respond to ever more demanding consumers.

And while the benefits can be obvious, the risk may be more hidden. Changes to IT architecture such as a move to cloud computing take careful planning. Tracking and managing the increasing number of digital assets (privately owned as well as owned by the corporation) is extremely hard. And there is a need to understand that digitisation that creates an advantage here may well create a problem elsewhere. All this is hard to come to terms with. Leadership needs to be strong, setting viable objectives, looking to the mid-term and the long-term, ensuring integration across departments as well as adequate resources, and always addressing people-centred issues. The benefits are huge but not easily grasped.

There are few organisations today that have not digitised at least some of their existing operations. Accounting and payroll functions have been computerised in some organisations since the 1940s. However, it was only with the launch of VisiCalc, the first spreadsheet software, in 1978, that financial modelling became possible.

Computers started to become common in offices from the early 1980s. Since then, many other parts of organisations have benefitted from digital technology.

Obviously, digitisation isn't new. However, it is becoming ever more powerful. That's obvious. So why are we writing about digitising operations now, when it is already happening? Because the opportunities continue to expand, continue not to be delivered well and change in ways that are hard to predict, for instance:

- Connectivity is enabling mobile and collaborative working in ways that are impacting business strategy, employee management and product development. Better connectivity allows greater quantities of information to be passed more reliably and more quickly (no waiting for files to download) between machines, irrespective of where they are located, making it easier for people at, say, the far end of a warehouse to have access to the data they need.
- Big Data analysis and artificial intelligence are helping organisations from hospitals to factories identify trends, predict and diagnose problems, and select solutions.
- Robots and cobots (collaborative robots – robots designed to work closely with humans) are not just making manufacturing safer and cleaner but they are helping office workers, farmers, builders, retailers, law enforcement, surgeons and restaurant staff (among others) become more productive.
- Computer-aided and driven production is giving more flexibility and less costly reset time.

We explore the detail of these and other technologies in Chapter 14. In this chapter we consider how understanding what emerging technology can mean for your organisation, as well as the risks around it, is essential for organisational leaders.

DIGITISATION OR DIGITALISATION?

Digitisation is the process of converting information into a digital form so that it can be manipulated by a computer with the intention of increasing the efficiency with which the information can be handled.

A different definition of digitisation misses out on that intention. Some people say that digitisation is simply about changing information from one type (analogue) to another type (digital). Any increase in efficiency to business processes, they say, should be called "digitalisation".

But that's an ugly word. And anyway, there surely has to be a purpose to digitising information, otherwise why would anyone do it? So, in this book we talk about the *digitisation* of business processes.

Understanding digitisation

Process digitisation allows organisations to move from slow, inefficient, paper-based – and typically manual – systems to fast, streamlined, digital workflows. This obviously makes sense from a cost perspective: organisations spend 5% and more of employee time on administrative tasks that don't generate revenue.

But the benefits go much further than cost and include: improved decision making as data is better collected and shared; better motivated employees who can be freed from boring tasks such as data input and frustrating tasks such as searching for information; better customer service as employees have access to more comprehensive information; improved quality as process and output monitoring becomes easier; and faster time to market.

Technology drivers

Digital technology is driving the enhancement of business processes because it is achieving four things:

- **Quality:** Technological improvements are resulting in devices that are more sensitive, smaller, cheaper to manufacture and more robust.

- **Connections:** Digital technology enables a machine to be connected to many other machines over the Internet of Things (IoT, see Glossary and Chapter 14); for instance, sensors that can connect more reliably to other machines using very little power or without a physical connection.
- **Knowledge:** Digital technology can be used to extract value from new knowledge that is created by combining and analysing information from different sources.
- **Autonomy:** Increasingly, systems and individual machines become autonomous and, driven by machine learning, increasingly effective.

Digitising processes

Digitisation isn't just something that happens in environments like factories and finance departments. It is happening across organisations, everywhere from board rooms to hotel reception areas. Here are just some of the areas of an organisation that digital technology can impact:

Decision making: Decision making can be enhanced by advanced analytics that will identify trends or likely outcomes. In addition, data can be analysed instantly, in real time, meaning that decision makers have the most up-to-date information available to them when they make a decision. At present many boards seem reluctant to pose questions to digital data sets during meetings, rather relying on limited, static sets of pre-prepared summaries or analysis created by others (often then aggrandised as Key Performance Indicators – KPIs) to inform their decision making (1). But, as new business leaders emerge, who are more familiar with handling data, this will no doubt change.

Data analysis on its own is often insufficient. The data together with any analysis need to be displayed in a way that will assist decision making, and also help convince others that the right decision has been made. There are many digital data visualisation tools that make it easy to interact with the data, sorting it in different ways or looking at subsets of data (such as last summer's sales vs this summer's sales). The tools also make it possible to create visualisations of the data very quickly and generally in very engaging ways, for instance, charts that appear to move by building automatically, moving over time, or showing different views.

Administration and record keeping: Paper trails, records of what people have done, are important across any organisation. Administrative processes already benefit from digitisation, with less space being needed for paper records and with faster and more accurate copying of data from one system to another.

Asset tracking: Digital technology makes it easier to track and manage information assets, for instance by including digital watermarks on them; unfortunately, they also make the creation of information assets (such as websites and social media accounts) easier, resulting in the need for better asset tracking.

Customer contact points: Receptions, call centres and anywhere that your organisation come into direct contact with customers and other stakeholders benefit from the automatic recognition of visitors or telephone callers, allowing people to be recognised automatically and staff to be given the information necessary to attend to them both more efficiently and effectively.

Meetings and collaboration: Digital technology already allows you to have "virtual" video-conference meetings where participants are not physically present but it won't be long before we can project a holographic image of ourselves into an empty chair that could either be physical or itself virtual. Virtual meetings save on travel costs but it can be hard to keep remote participants engaged (it's hard to see if they are checking their emails while you are talking) and virtual participants can find they are less able to get heard and thus influence discussions.

Digital technology goes a long way beyond virtual meetings though. It can allow businesses to communicate and share information easily. Landlines can be integrated with mobile phones and Wi-Fi can be integrated with 4G, 5G and fixed Ethernet connections. Enabling different technologies and devices to communicate seamlessly (called Unified Communications) enables better customer service. And it allows more efficient business processes including the ability of teams, whether physically together or distributed around the world, to work together on documents, designs, presentations and prototypes.

Taken to an extreme, digital technology might be seen as meaning that there is no longer a need for physical business locations, except where people are required to handle goods or machinery, or interact with customers. This saves money and commuting time. However, having a fixed place of work is often very important to people. Travelling to it adds structure to the day. It is somewhere where they can interact with colleagues: home-workers often feel very isolated as well as powerless as they are unable to interact physically with managers. And importantly, it can provide them with a personal space (which is why many people dislike hot desk environments where they don't have a place to put potted plants and family photos).

Innovation: The days of the (paper-based) suggestion box are over. Forward-thinking companies use digital innovation platforms such as Wazoku.com to collect ideas, prioritise them through employee voting and other forms of evaluation, and enable the owners of ideas to track them through the acceptance and development process, providing valuable feedback that encourages their further ideas.

Product lifecycle management: Challenged by a period of rapid expansion, clothing manufacturer SPANX employed a digitised product lifecycle management tool that enabled them to shorten new product development, react to market changes more rapidly, support a wider set of products and service their retail partners (and end consumers) more effectively. As a result, marketing, design and manufacturing are far more closely connected, with manufacturers who are able to get product information they need, such as design specifications, measurements and production notes, at the touch of a button.

Training: While computer-based training isn't on its own going to fulfil the training requirements of any organisation, it does have a number of benefits such as: the automatic recording and analysis of test results; automated communication with managers when training is, or isn't, completed; the ability to add a competitive element such as leader-boards; and the ability of employees to access the training anywhere and at any time.

Manufacturing: In a digitised factory, machines are connected with people, their devices, the wider organisation and external stakeholders such as suppliers. The main purpose is to achieve increased efficiency to boost margins or deliver additional services for the same cost. But digitising operations will also allow for increased variations and customisation of existing products. In addition, it can allow the progress of products along production lines to be tracked.

Factories also need to consider the potential of new and emerging technologies such as new smart materials (for instance materials that can conduct data or are self-healing), nano-machines (very small machines) and 3D printing. GE is using 3D printing (or "additive manufacturing") to manufacture some parts for its engines as and when they are needed. Parts include fuel nozzles that previously had to be assembled from 19 separate parts. These 3D printed parts are five times stronger than nozzles manufactured traditionally.

Moving to the cloud

One of the major process changes at present is the move to cloud computing.

Cloud computing involves storing and accessing data and software over the internet rather than on computers held in your workplaces. The move to the cloud is sometimes managed as a major IT project. But often cloud computing services are used privately or by an individual within an organisation. Anyone who is using Dropbox to store and share files, for instance, perhaps with a client or a supplier, is using the cloud.

One common example of cloud computing is when organisations store their data in the cloud. They may do this for a number of reasons:

- It's cheaper. Instead of buying your own computer to store your data you use part of a larger computer that in effect you are sharing with other people, which can keep the cost down.
- It's more flexible. As your storage requirements go up and down you can flex your requirements, only paying for as much as you need.
- It's safer. Cloud computer businesses are specialists in what they do and one of their basic hygiene factors is the ability to keep data safe. Because they are specialists, they are probably better at keeping data safe from hackers than you are.
- It's reliable. Cloud computer businesses have to provide very high levels of access for their customers if they are going to stay in business. So they have good physical security and power backups. And they may well store your data in more than one place so that if one of their facilities has a problem you still have access to your data.

But of course there are always downsides! And for storing data in the cloud these include:

- You need access to the internet to get access to your data. If your organisation has a problem with its internet connection, then it may be unable to perform its tasks.
- By using a third-party cloud-computing supplier you are having to trust them with your data; and while they are probably better at keeping it safe that you would be, this is still a risk, especially if your cloud services provider is also working with other companies including perhaps your competitors.
- You are also having to trust that your cloud computing supplier will always be able to make your data available. But even the largest companies can have problems. For instance, an outage on Amazon Web Services in 2017 cost companies $150 million (2).
- Flexibility. Due to constraints in service agreements or other practical problems, it may be impossible to select and delete individual data files from a cloud service, especially from backs-ups of the data; this could potentially cause problems if there were to be a legal requirement to delete certain data (for instance as a result of a GDPR "Right to erasure" request).
- Latency. Because you are accessing your data via the internet there may be a delay between requesting something and getting it. This can cause frustration for end users.
- Your cloud service may involve sharing a computer with another organisation and in rare cases this has resulted in data from one organisation becoming available to another organisation.
- You are reliant on a third party. If you have a dispute with them or if they go bankrupt it may be difficult to get your data back. And if you want to move to a different vendor it may be difficult to do this because of difficulties migrating from one platform to another.
- It may be difficult to track exactly where your data is being stored; while cloud computing services may promise to store your data in a particular place or country (which may be a requirement of privacy regulations if you are storing personal data), the backups (and the backups of those backups) may be stored elsewhere.

Most of these disadvantages can be managed, at least to an extent. And on the whole, cloud computing services have a lot of advantages, which is why so many organisations are using them (3).

Making the case for process digitisation

The case for digitising business processes comes down to two words: efficiency and data.

Efficiency and waste reduction

Is your organisation using digital technology to eliminate as much waste as possible? Eliminating waste is one of the most effective ways to increase profitability. This fact was recognised by Toyota's Chief Engineer Taiichi Ohno in Japan who categorised unnecessary waste in manufacturing (muda in Japanese) into seven areas. Solving these seven areas of waste has become known as "lean manufacturing".

It is possible to use the ideas behind lean manufacturing to address waste in any organisation through more efficient digital processes. Let's take a look (Table 4.1).

Table 4.1

Manufacturing waste (muda)	Business process waste	Digital technology solution
Overproduction of goods	Taking unnecessary action or action before it is required	Automated processes, especially using artificial intelligence
Waiting for raw materials to be processed	Waiting for a colleague to finish something; project and process dependencies	White-boarding and other collaborative software; the use of Optical Character Recognition (OCR) to transcribe paper documents
Transporting finished goods unnecessarily	Moving people around unnecessarily	Video conferencing; remote and mobile working
Inappropriate processing	Scope creep; over delivery; inappropriate quality levels	Templated processes that limit scope for over-delivery
Unnecessary inventory	Too many paperclips; too many people	Document digitisation and storage; automation
Unnecessary movement of components	Unnecessary sign off stages; delays due to sharing information	Document analysis and automated validation
Defects in products	Human error, e.g. in data transcription or storage	OCR, automated document checking

---

MAKING PROCUREMENT MORE EFFICIENT

There is at least one example of a school in the UK that, seeking to exert more effective financial control, has created an overly complex procurement process, building in unnecessary processing. When procuring anything, including such staples as pencils and exercise books, every step in the purchase order process (need to purchase, draft purchase order, actual purchase order, receipt note, invoice receipt, and finally, payment) needs sign off by the Head Teacher.

This is arguably an entirely excessive amount of control. However, it could be made much more bearable through a digitised process where the complex paperwork was integrated and where the items for approval were queued and then signed off in batches.

Efficiency and productivity

But efficiency isn't just about decreasing waste. Digital technology can help in other areas too. It can help you stop losing things: tools and materials can be tagged so that they are easily located or so that they can come with additional information if required. It can help you find things: using RFID communication technology (Radio-Frequency IDentification, a technology where data encoded in RFID tags or labels can be accessed and read by someone a short distance away), people can find information such as work instructions, storage instructions, stock level instructions or tracking numbers. And human error can be reduced if data transcription from one document to another takes place through a digital system.

Factories in particular can be made smarter. Sensors on machines can signal when maintenance is likely to be needed or when a fault is developing (predictive maintenance). Prototype parts can be created using augmented reality to create a virtual design and then additive manufacturing to create an inexpensive physical prototype which can be tested for fit before the part is machined. Autonomous machines (robots) can take over dull, dangerous and dirty jobs, leaving human workers to contribute greater value. Wearable technology can be used to monitor the health and attentiveness of machine operators.

Data and information

The inevitable outcome of digitising a process is the production of (yet more) data. Not all that data will be useful and it's important to avoid drowning in data when you digitise processes. But at least some of that new data is likely to be of use in various parts of the organisation, providing insight into market trends, customer profiles, operational deficiencies, potential risks, etc.

We talk more about using data in other parts of this book. But at this point it's worth emphasising that pure data isn't much use and that information and knowledge are likely to be more valuable (Table 4.2).

All business processes yield data all of the time. The figures on an old-fashioned till roll in a café are data, for instance, data that before the days of computers would have been transcribed by hand into an accounting ledger. In the same way, 25 years ago, customer complaints would have been written down in a log and the response to them also logged. But with a digitised process, data can be collected automatically and, in some cases, used to drive automated decisions at the front-end on the specific cases, whilst providing trend and performance data out of the back-end at the same time.

Digitisation drives the production of additional data. And organisations need to stay on top of that, deciding what data is useful on its own, useful if combined with other data, or worthless. The reality is that quite a lot of data is going to be worthless, or at least not worth processing. The café's till roll data might tell you that the tables by the window are more popular

Table 4.2 The path to wisdom

Data	Symbols (e.g. words and numbers) that are largely meaningless on their own	Two. Elephant. Garden
Information	Data that has been processed or given structure and context so that it has meaning to a human	There are two elephants in the field near my garden
Knowledge	A human's experience of what information means	Elephants are heavy and will damage my dahlias if they trample them
Wisdom	Knowledge from one area synthesised with knowledge from another to give insight into the future	If I shout and wave, I will scare the elephants away like I did with the rhinos

but the waiting staff could probably have told you that without the need for expensive data processing.

Data is useful, but generally it needs to be combined with other data or contextual information if it is going to help you make decisions. In other words, collecting data as a result of digitisation shouldn't be an end in itself, especially as the storage of data can represent a risk to your organisation, especially if that data includes personal data (see Chapter 12) and it costs large amounts of money to store it. Rather, it should be something that fuels a process of generating useful information that the business can then exploit.

Advantages beyond efficiency and data

Digital technology brings a number of other benefits beyond efficiency. Here are a few examples:

- It can increase information security: It's far easier to keep confidential documents secure in a computer than a cabinet (that's not to say that computers are always secure though, as we discuss later in Chapter 12).
- It can help you meet the expectations of younger employees or customers who may assume that digital technology is appropriate (even if it delivers little of value, which may sometimes be the case).
- It can help you to provide better services and products such as enabling customers to track when a parcel is to be delivered (we talk about this more in the next chapter).

Issues with digitisation

Given the very substantial benefits of digitisation, it seems odd that anyone should argue against it. But it's not always appropriate. As with so many other things in life, just because you can do it does not automatically mean you should do it (or do all of it).

When not to digitise

Real people are often an essential part of a process. For instance, when you visit a high-end restaurant or a hotel you probably want to interact with a human rather than an ordering screen as you might in a fast-food venue. And if you are meeting a business colleague for the first time it is generally better to do this face to face than via a virtual meeting, not least so that you can build a more social relationship with them afterwards.

There are a number of other situations when human-to-human contact may be beneficial in ways that a human-to-computer process isn't, such as picking up on body language or having a conversation where a record isn't kept.

On other occasions digital technology can simply be ineffective. Forcing designers to use digital technology to straight away go into detailed design, when many will prefer to start off with a sketch that is hand-drawn on paper, is not sensible. But you could provide them with a tablet computer to capture that sketch and to then enable further processing of it.

It's also important to remember that people are (at least currently) better than computers at some things including the recognition of exploitable information, e.g. the useful patterns in data sets rather than just patterns. For instance, a very detailed analysis of listeners to a radio station might find that a majority were left handed: interesting information but useless to anyone wanting to maximise advertising sales on the station, unless they have a client who wants to target left-handed people.

And of course, there will be many instances where the business case for digitisation has been incorrectly made. The risk of unnecessary expenditure and disruption around process digitisation that shouldn't be happening is not small. Do you really need an electronic visitors' book when a simple paper system would do the job just as well?

Related to this risk is the temptation to reinvent the digital wheel when there are good off-the-shelf digital products available. We know of at least two media companies that felt that they needed to create their own online content management system from scratch when there were plenty of well-established off-the-shelf products available.

Why does this happen? Perhaps because of ego – "We are special and so need to develop our own solution". Or perhaps simply because of a lack of knowledge about existing ready-made solutions.

It is important for you to bear in mind Pareto's 80% rule. In this case an off-the-shelf solution, delivering 80% of requirements with the balance of functionality being done manually, might well be much more cost effective and quicker to implement than building a content management system from scratch – even one that is 100% perfect (which in reality, if you are starting from scratch, it probably won't be!).

That's not to say that building your own software solutions is never appropriate. When you have very particular requirements that would mean that any commercially available software would have to be heavily customised, and therefore hard to maintain over time, it may well be better to start from scratch. But a solid business case must be made to support whatever decision is taken.

Starting from scratch

When digitising systems, organisations have the opportunity to buy in a ready-made (and well-tested solution) or to build one from scratch. This is an important choice to make. Starting from scratch and building a new system from the ground up can mean that you end up with a system that delivers absolutely everything you need. Or it can mean that you end up with an expensively built imitation of something already available on the open market, but an imitation that has not been so well tested over time.

Because of the risks of building a new system from scratch, many organisations decide that they will instead buy in established technology from a third-party supplier and customise it to their requirements. But customising systems takes time and trouble. The customisations required needs to be identified and prioritised. When the requirements are implemented, they need to be tested and recorded properly so that when the basic system is updated the customisations can be updated too.

One of the UK's best performing modern (20th century) universities decided to implement a new student management system. It then selected an unproven system from the USA and decided that it then needed to substantially customise it using a very small team who also had day jobs.

Unsurprisingly, a post-implementation review by consultants castigated the organisation's governing body and top management who, collectively, had shown little interest in a fundamental organisational component or in the over-worked staff trying to deliver it.

Whether it is better to build a new system or customise an existing system will depend on the complexity of your requirements. There is no right or wrong answer.

Watching out for the risks

When you are deciding whether to digitise any of your processes, you need to remember that digital technology may well introduce some risks. The management of the risks of process digitisation may be difficult if there is a lack of experience in the organisation.

Why are digital projects so risky? There are a number of common reasons:

- The risks they contain may not be well understood if they haven't been experienced by anyone on the team.
- The skills required to manage the risks may not be available in-house, so they may involve a long-term dependency upon third parties, such as contractors or agencies, who are more loyal to themselves than to you.
- As they are by definition new projects, they may be experimental with fuzzy goals and outputs that have not been defined because they are hard to visualise.
- They may focus too much on the technology, merely adding a computer to the process rather than digitising the process.
- They are often driven at speed – by enthusiastic developers who want to ignore process (or compliance) and get to a product as soon as possible.
- They are often focussed on an endpoint (the product) rather than continued business success.
- People involved with them may use jargon or words that mean different things to different people meaning that misunderstandings and disappointments can easily arise.
- Because technology changes so rapidly they may be obsolete before they are finished (that's no excuse for not starting them, however).

Managing the digital estate

The digital estate comprises all the digital technology – software and hardware – that exists in your organisation. And it does not exist in a vacuum. It's not just a collection of individual word processors, telephones and calculators. Rather it is an ever-changing network of interconnected machines that makes up an intrinsic and increasingly critical dimension of any business. How this network is acquired, managed and disposed of will have a major effect on the degree to which your organisation thrives.

We aren't going to talk about procurement processes here, or how to get a bargain. While it is important that technology assets deliver good value, it is for IT professionals to decide on what assets to buy, how much to pay for them and how to integrate them into the organisation as part of the discipline of IT governance.

Rather we want to paint a picture of how organisations need to have the right philosophy when it comes to investing in, managing and tracking their digital assets. (It would be perfectly reasonable to say this too is part of the realm of IT governance, but it is a part that you need to be very much involved with, which is why we cover it here.)

The digital estate will always be a mixture of digital machines including:

- Individual devices such as desktop computers, computer accessories such as headsets and keyboards, telephones, mobile device chargers, printers and computers used to store data.
- Connections that connect individual devices to one another and to the outside world – such as network cables, internet routers, Wi-Fi boosters and servers.
- Software that enables the individual devices and the humans that use them to perform tasks.

These digital assets should generally be treated as being part of a network, rather than individual devices. And because they are part of a network, any individual machine (including software) needs to be considered alongside the other machines in the network.

There is another set of machines that can be considered part of the digital estate: machines that have computers in them but that are not principally thought of as computers; these could include factory robots, autonomous drones, security cameras, even motor cars. These machines will also be part of the digital network and can affect its security.

It is important to realise that the digital technology owned by an organisation isn't the only technology that will be operated within it. People often bring their own digital devices (smartphones, tablets, laptops) to work, for their own convenience as much as the organisation's. Managing the Bring Your Own Device (BYOD) culture is important and decisions need to be taken as to how far it is allowed to go, decisions that in large part will depend on your organisation's cyber security profile. We discuss BYOD and related issues in Chapter 11.

Recording the digital estate

In order to create an efficient technology environment, we need to know what machines are part of it. It might seem simple to keep track of all the digital assets in an organisation. But for a large organisation, this can in fact be a remarkably difficult thing to achieve. This is because there is a need to know:

- The number of machines (desktops, laptops, telephones, printers…)
- The make and model of each machine and whether it has been modified in any way
- The physical location of each machine
- The physical condition of each machine and whether it is under warranty
- The connections and relationships between them.

There is also a need to know about software on the network:

- The software programmes on each machine including its version, whether and how the software has been configured, and, critically, whether we have a licence to use it and what the licence number is
- Other software on the network such as digital images, documents and websites
- If software has been developed or customised internally there will be a need for adequate records of how it was developed so that it can be maintained.

Managing these assets is important as it will enable you to:

- Avoid duplicated purchases or maintenance of unnecessary software licences for computers that are no longer in use
- Identify opportunities to increase efficiency by replacing outdated machines or software

- Protect the network's efficiency and security by ensuring software on machines is up to date and configured to certain standards

But this can only be done if there is a record of all the assets. And creating an accurate record is a difficult job for any but the smallest organisations. As the digital estate of most organisations has grown substantially, and often in an unplanned manner, over the last 20 years, complete information may simply be unavailable. And if it is available it may well be inaccurate. In addition, the Internet of Things is making the creation of an accurate record of assets even more difficult as more and more devices – coffee machines, printers, thermostats and many more – arrive ready to be connected to the internet via office networks.

The theoretical solution to this problem would be to audit (count and examine) all the devices. But this is likely to be impossibly time consuming as well as disruptive. For many organisations the only practical solution is to turn over a new page and start doing better going forward. Where possible, a progressive deployment of a configuration management database (CMDB) will be beneficial to all of the above.

Owning vs renting assets

Increasingly organisations have a choice whether to own or rent their digital assets. Putting money aside, the main benefit of renting is keeping up to date.

Many software rental contracts allow for rented software to be updated on a regular basis. An example would be the Microsoft Office suite of software (Word, Excel, PowerPoint, etc). This software, launched in 1998, was originally sold as a one-off purchase but in 2011 Microsoft offered a subscription version that allows people to pay for access to a continually upgraded version – "software as a service" (SaaS) – that they would access over the internet.

Benefits of SaaS include:

- Users can access the software on any device they choose to via an internet browser, without having to download and configure software.
- Users always have access to the latest version which, as well as having the latest functionality, is also likely to be more secure against hackers.

- Organisations can scale their use of the software as more people need it, benefiting from flexible purchasing.

One potential downside of SaaS is that in certain environments regular updates of software may be unwelcome. IT professionals may want to test new versions of software to ensure they don't cause problems, for instance with machinery. This is one of the reasons that the WannaCry computer virus caused problems in the UK's NHS and a number of large organisations in May 2017 (see the text box in Chapter 11).

Digitising internal processes is only half of the opportunity for most organisations. There is also significant benefit in considering how to digitise products and services, as we will see in the next chapter.

References

1. Dibb, S. *et al.* (2019). *Digitisation and Decision Making in the Boardroom | www.nemode.ac.uk*. Guildford: Surrey Business School. [online] Available at: www.nemode.ac.uk/?page_id=1524 [Accessed 24 May 2019].
2. Hersher, R. (2019). *Amazon and the 150-million Typo*. [online] Available at: www.npr.org/sections/thetwo-way/2017/03/03/518322734/amazon-and-the-150-million-typo [Accessed 24 May 2019].
3. The UK Government has a "Cloud First" policy even if it is not always observed. Its guidance can be found at: GDS. (2019). *Government Cloud First policy*. [online] Available at: www.gov.uk/guidance/government-cloud-first-policy [Accessed 24 May 2019].

5

TRANSFORMING PRODUCTS
AND SERVICES

Summary

Digital technology offers the opportunity to transform products and services. In some cases, such as media and entertainment, organisations have had little option but to engage with digital technology as it radically changed, and continues to change, their industry. But in many other industries the digitisation of products and services is still, to some degree, an option, albeit one that needs to be taken very seriously.

Digitisation can mean enhancing existing products with add-ons such as additional information. It can mean changing a product into a digital format. It can even mean developing a new business model that is enabled by technology. All of these changes can have a radically positive effect on organisations if they are managed properly and if organisations develop the vision to imagine what additional value their future products and services might deliver through digital technology.

In the last chapter we talked about digitising existing processes and the new flexibilities that can bring. But digital technology doesn't just offer a chance to reduce costs by digitising internal processes. In this chapter we examine how organisations can invest in digital technology to more completely transform the range, variety and nature of the products and services they offer to their customers.

The transformational opportunity isn't about simply saving some money. It's about delighting the customer with better services that are improved through faster (sometimes automated) decisions, more streamlined processes such as shipping and more intensive communication between company and customer.

Models of digital transformation

Digital transformation really means one of three things:

1. **Digitally enhanced products and services:** This can involve adding a digital element to a physical product. An example might be a motor car which has a digital satellite navigation incorporated. It could also involve using digital technology to create a wider range of products or to create a more customised product.
2. **Digitised products and services:** This involves turning the whole product into a digital format. As an example, a magazine that used to be produced with ink on paper can be produced, often significantly more cheaply and made available to a wider audience, online.
3. **New business models:** This involves using digital technology to enable a whole new way of offering a service, for instance renting out goods when previously you had sold them outright. It can also involve taking an existing business model to a new level or creating a more vertically integrated service.

Digitally enhanced products

Digital transformation most commonly means adding a digital element to a physical product in order to provide better value to your customers. Not all digital additions are truly *transformative* transformations of course. In fact, most are not. But they fit within our definition of digital transformation: using digital technology to change a product or service and/or the customer's experience of it.

Adapting to digital services

There is nothing new about providing digital services. In the UK, supermarkets have been providing online ordering and home delivery successfully since the turn of the millennium. And banks have been providing High Street cash machines (ATMs) since the late 1960s. The issue is how you adapt to provide such services.

The move to online shopping caused supermarkets a headache. There were costs incurred with delivery (this was soon charged for) and basket sizes tended to be smaller. Supermarkets reacted in a dozen different ways: enhancing loyalty programmes, sending out customised coupons, increasing the number of free samples, enabling shoppers to save their shopping lists from one shop to the next, even encouraging shoppers to choose the items they wanted discounts on. Online grocery shopping continues to grow: in the UK around 50% of consumers do at least some of their grocery shopping online. But it is equally important for you to realise, when establishing your digital strategy, that 50% still don't.

Banks faced a different problem when ATMs became commonplace. No longer was there such a need for customers to visit a local branch to transact. While this saved the cost of employing so many cashiers, it reduced the opportunity that banks had to upsell their customers with loans and insurance products. Some banks have addressed this issue by attempting to turn their branches into "destinations" for shoppers. Automated banking through cash machines is still available in these branches but an attempt is made to hold customers physically. Metro Bank in the UK encourages customers in with their "magic money machines" where children can turn pocket money coins into notes, doggy treats, sofas and a generally welcoming atmosphere. Capital One bank in the USA takes this a step further, positioning itself as a coffee shop with ATMs and computer terminals. It's still a bank, but one where the relatively few staff have a chance to meet their customers face to face.

Customisation as the norm

Digital technology makes mass customisation easy. Manufacturers can show a limited set of options to customers, and customers can select from those options. The concept is well established in many industries including travel, automotive, desktop computers and food (mymuesli.com allows you to mix your own breakfast cereal with a selection of fruits, nuts and chocolate).

Mass customisation is underpinned by digital technology that allows the consumer to communicate their selections to the manufacturer. But much more significantly, the manufacturing process is automated by digital technology allowing it to cope with far more combinations, meaning that each item can be built to order but still at a competitive price.

Personalisation

The digital world makes it easy for a customer to send complex information to a company they are buying from. Limited personalisation is quite common in the fashion industry: Nike, for instance, offer customers the chance to add their initials or a motto to their trainers. Unilever lets you order a jar of Marmite with your name on the label. (It doesn't taste any better – but then, how could it?) Other companies such as Moonpig and Truprint allow you to upload your own photos to provide truly individual products, such as greetings cards and mugs.

Both mass customisation and personalisation have some downsides. Cost is one of course. Factory automation requires investment, meaning that consumer prices are likely to be higher than for mass-produced items so these are strategies more appropriate for premium products. In addition, handling returns is more difficult, both for mass customised and personalised goods. And there can be issues with managing the additional suppliers that may be necessary to include in the supply chain.

Providing information

It's relatively simple to provide consumers with more information about your products. With food retailing especially, this is becoming more and more important, whether people are researching their shopping online or shopping in the High Street. In China, the Hema supermarkets provide customers with product information including sourcing, brand heritage, price points and nutritional value, all accessed via a QR code on the item and a handy mobile phone app.

Informationalisation

Providing straightforward product information is a hygiene factor: many consumers want this information, but it doesn't enhance the product. In contrast, informationalisation (admittedly, a horrible word, it means making a product more "informational" or full of information) involves the addition to a product of information that adds value for the consumer.

A famous example is Coors beer where the mountains illustrated on the can turn blue when the beer is cold enough to drink. A more digital example is a car that informs the owner when it needs it to be serviced or when a tyre is under-inflated.

It's relatively easy to find out what information consumers want. But it will be a lot harder to uncover how to enhance your products and services with extra information. The first step though is to be open to the idea that there is likely to be some value that you can uncover and that will give you a competitive advantage.

Being digital

Being digital (as we said earlier, some people would call this digitalisation) involves adding a new digital service (as opposed to the providing of additional information) that enhances the physical product. Building a satellite navigation system into a motor car is an example of this: the car would work perfectly well without it but the technology adds value for consumers.

Another form of being digital is the replacement for consumers of an important real-world process with a digital process. One example here is the development of a totally online mortgage application by UK bank Nat West. This is a very powerful tool for businesses because as well as providing a consumer benefit, it may well be reducing costs.

Sticky ecosystems

There's nothing new about trying to lock your customers in. And digital technology makes this a bit easier. Apple has made all of its products (MP3 players, phones, computers, tablets, wearables) highly compatible. If you buy one Apple product there is a whole family of other products you can buy that you can be confident will work well with it. It's easy, for instance, to transfer music from an Apple computer to an Apple phone. The products, however, don't work so well with other manufacturers' products. Once you are in the Apple ecosystem it's hard to get out!

Digitised products

In some cases, organisations have gone beyond enhancing their products. These have moved from physical products to digital products, creating digital look-alikes or new digital-only products, sometimes for totally new markets.

From atoms to pixels

Many media companies – especially broadcasters and newspapers – have been forced to create pure digital (online) products in order to keep competitive. The largely downward spiral for traditional media companies has been caused by the internet, which has destroyed some traditional revenue streams (classified advertising) and reduced the value of others (brand advertising).

Online it is relatively easy for anyone to set themselves up as a global media company. There is no need to buy expensive printing presses or broadcasting equipment. A video camera, a laptop, an internet router and you are all set to be the next Rupert Murdoch.

The woes of the media industry have had plenty of attention elsewhere. But there are analogous learnings to be had, learnings that will be relevant to industries well beyond the media:

Cost of technology: Are you developing new products that use technology that has fallen in cost, or are you stuck in a mindset that means you refuse to abandon obsolete technology because it cost you a great deal only a few years ago?

Competition: The nature of the internet means that it is easy for new competitors to enter the market and talk to a global audience, substituting their product for yours. How will you reposition your organisation and your services to avoid having a brash new competitor eat your lunch in your home market?

Immediacy: The internet allows information to circle the globe almost at the speed of light. Are you able to share your information with your audience as quickly as they have come to expect it?

Direct access: The internet gives you direct access to consumers, without the need for any intermediaries. While this isn't always a blessing, there are certainly advantages that can come from it including greater profits (as disintermediation reduces the need to pay third parties), the ability to collect data about consumers and the ability to communicate with them as individuals more effectively.

Functionality: Are you taking full advantage of the interactive functionality the internet provides to exchange information with your target audience as a way of engaging them?

Reusing infrastructure

Going digital means an investment in technology. And as with any investment, when you maximise the use of that investment you maximise results.

Amazon's story illustrates this. The company started out as an online bookshop but quickly realised that their infrastructure could be used to sell things other than books. The company then went a stage further, renting out its own IT infrastructure to other companies. And now, Amazon Web Services is one of the largest IT suppliers in the world.

Reusing data

While reusing infrastructure can be very profitable, reselling data collected in the course of your operations can be even more so.

Have you got data that you could sell to your existing customers? Agrochemicals giant Cargill has developed a new service to supplement its business of selling seeds to farmers. Cargill collects a great deal of data from its customers and has collected a mass of data on how its seeds perform in different types of soil and under different weather conditions. Analysing this data enables it to sell customised advice to individual farmers.

Rolls Royce's aero engines are used the world over in many thousands of planes. But their services go much further. Rolls Royce essentially, and most often now, sells a service package of the engine, remote but immediate data collection and analysis of its operational performance and pro-active and anticipatory maintenance. This gives Rolls Royce improved incomes and increased customer dependence in return for reduced total costs of operation (TCO) for its customers. This win-win is entirely dependent upon digital technology that was truly transformative when Rolls Royce pioneered it.

But you don't have to restrict your data sales to existing customers. Numerous companies have discovered totally new markets for their data. BMW, for example, takes telemetry readings from cars and uses it for its CarData marketplace. Garages, insurance companies and fleet operators can buy access to this valuable data. In a similar way, Toyota has built a new business that uses the satellite navigation devices it installs in cars to capture the speed and position of cars. Traffic data can be sold to governments as well as to corporate fleet operators.

Away from the automotive industry, UnitedHealth built a $5 billion business by analysing the information contained in the many claim forms it processes. Pharmaceutical companies analyse this data to discover how their drugs are used, how effective they are, and how well they compete with rival drugs. This new and highly profitable business, Optum Insight, delivers a growing revenue stream that UnitedHealth would have missed completely if it had focused purely on its core pharmaceutical business.

Sea freight line Stena takes a slightly different approach. They collect data about tides, currents and the depth of the ocean as their ships sail around the globe. Working in partnership with Hitachi, the data is analysed and made available to anyone who could benefit from it, enabling other freight carriers to plan energy-efficient shipping routes.

It could be argued that Stena are losing revenue or even eroding their competitive advantage by giving this data away. And that is always going to be a consideration. For Stena and Hitachi, the lost revenue and competitive advantage are balanced out by the wider benefits to the environment. But that won't be the case for every company with valuable operational data to use.

Extrapolating data

Data can be processed to create new data. With the costs of data processing falling seemingly by the day, it is increasingly possible to extrapolate additional data from the data you have. For instance, the customers of car manufacturers are often offered a configurator so that they can customise their car. As well as allowing the customer to specify the car they want, this process delivers a huge amount of data to the manufacturer. What options are never chosen? Are certain options more popular at certain times of the year or in different parts of the country? Do people often try to choose a combination of options that are not available? This data can be used to optimise the car manufacturer's products and their marketing, as well as potentially offering routes to associated reductions in operating costs.

New business models

The reuse of operational data can provide organisations with a new market. But in some cases, there are entirely new business models that can be enabled by digital technology, and especially by the communication and data exchange that digital technology allows.

Perhaps it is rare for established businesses to disrupt industries by developing new business models. Kodak (1) were famously reluctant to compete in the digital camera market that they had helped to create. In the same way, Sony were slow to digitise their Walkman personal music range, allowing Apple's iPod to steal much of the market.

There seem to be several reasons for this but two stand out: a reluctance to reduce or limit existing sales for the sake of future sales; and lower digital skill sets in traditional industries. Both of these are issues that merit board-level attention.

The sharing economy

Companies like accommodation broker Airbnb and parking broker JustPark enable private businesses and ordinary consumers to share their assets with other consumers. In a similar way, at least at present, taxi company Uber allows individual taxi drivers to work for themselves rather than for a minicab company. Flexibility for the driver, lower overheads for Uber – or so goes their story.

Usage-based pricing

It often makes sense to rent rather than buy. Companies like Zipcar can free consumers, especially in cities, from the costs and hassles of owning a car. In the construction sector, Hilti enable consumers and businesses to avoid upfront costs and asset management costs, instead paying only for the assets they use. And in printing company Heidelberg's digital business model, customers only pay for the number of sheets actually printed rather than buying printing presses and then paying additionally for consumables or services.

Limited free services

Another pricing model involves offering a service free to entice users in, and then use another, better service but at a cost. This is often known as "freemium" (a contraction of "free" and "premium"). The cloud storage and collaboration company Dropbox is a wonderful example of this with a free service that includes a limited amount of storage: after a while, the service can become unusable because the free storage has been used up and the user is then offered the option of paying for more storage. The business-focussed social media platform LinkedIn is of course another.

New obligations

A new business model may result in your business being bound by new rules as you enter a new marketplace. For example, car companies moving into finance, to help make more sales, were undoubtedly unaware of how much more compliance obligation they had taken on then. And that was nothing compared to what they now have to do. We explore compliance further in Chapter 10.

Make the customers do the work

It's always a good idea if you can get someone to work for you for free. Travel company TripAdvisor is an expert at this, encouraging its users to leave reviews that will help other travellers and prompt them to buy holidays.

Many other businesses suggest to customers who have called a helpline that, rather than waiting for an advisor, they can probably find the answer they are looking for online. And shops that allow the customer to check out themselves rather than using a cashier are saving the cost of employing a cashier (but risking additional shoplifting). Of course, there is nothing new about this: banks have been encouraging their customers to serve themselves at ATMs since the 1960s.

Transferring the costs of certain actions from your organisation to your customers is often a very effective idea. But to ensure that your customers remain happy with your services when you do this it may be necessary to persuade them in some way. In the next chapter we consider how digital technology can be used to communicate effectively with your customers, and how it needs to be managed so that your organisation's reputation is not accidentally damaged through your digital marketing efforts.

Reference

1. A useful analysis of Kodak's digital "misses" is given here: Pachal, P. (2019). *How Kodak Squandered Every Single Digital Opportunity It Had.* [online] Mashable. Available at: mashable.com/2012/01/20/kodak-digit al-missteps [Accessed 24 May 2019].

6

DIGITAL MARKETING AND SALES

Summary

One of the great benefits of digital technology is the way that it allows data to be created, collected and shared. And data about your customers is one of the most important instances of this. It's the potential of deriving and using new customer data to identify, convert and retain customers that drives the adoption of digital technology by marketing and sales departments.

Of course, the benefits of digital technology go a long way beyond customer insight. There are advertising channels like email, social media and websites to use. There are exciting new advertising formats such as augmented reality to experiment with. There are powerful ways of tracking prospects and consumers with the aim of making sales more frequently and at higher price points. But there are challenges too: managing marketing assets and employing the people skilled in their use isn't always easy. Leaders need to be aware of these challenges and ensure their organisations are meeting them.

In 2016, the business research company Gartner famously predicted that Marketing Directors would be spending more on digital technology than IT Directors by 2017. Whether or not this has proven true, this chapter looks at the particular opportunities and challenges that investments in digital technology pose for marketing and sales teams, and the oversight this increasingly large investment needs. And along the way we have tried to bust a few common myths.

Marketing and sales have always depended largely upon data. Their success is based on the quality of information that they can gather. Is it accurate, complete and up-to-date (or, if you like, is it the truth, the whole truth and nothing but the truth)? And that success is based on the ways they are able to exploit this information in order to deliver more competitive products and services (more attractive, cheaper or functionally better) and better customer service.

Their dependence on data is in large part why they are now investing so much in digital technology. It's data that identifies the most competitive products, the best target audiences, the most effective advertising messages. And this marketing data is best managed through digital technology.

Understanding consumers

Many, but by no means all, consumers are heavy users of digital technology. This will normally mean that they leave a large digital footprint that can be used to understand who they are and what motivates them. One problem with this is that consumers increasingly find this use of "their" data to be either intrusive or simply irritating, both of which can damage brands.

Nonetheless, digital technology offers a number of powerful ways of understanding consumers:

- **Social media activity**. The things that people write about your company and your industry on social media can increasingly be collected and analysed automatically, giving insights into what people like and dislike about your products, customer service, advertising and competitors. Whilst "sentiment analysis" of the comments that people post isn't perfect, it is a useful pointer to whether things are improving or deteriorating in the minds of (at least some of) your customers.

- **Website visits**. You may not know exactly who is coming to your website all the time, but you can see what they are looking at. A detailed examination of online behaviour, especially around product options you offer for sale, will show what is popular and what isn't. Examination of search queries on your website may add some further colour.
- **A/B testing online**. This can enable you to run numerous experiments on pricing, product options or marketing messages on small numbers of people, comparing version A with version B and identifying the version that performs better, before you bring the better options to a wider market.

All of these methods need to be used with care. Understanding the consumers who leave a digital footprint isn't the same as understanding all your consumers. Some may not be online at all (around 20% of UK adults rarely if ever go online). Others may not leave a relevant footprint. The behaviour of others online may not reflect their behaviour in real life. And your interpretation of their behaviour may be faulty.

Digital marketing myth 1. Everything that matters is digital

Not all marketing uses digital technology. There is nothing very digital about a full-page advertisement in *The Times* newspaper. And while the audience analysis that persuaded the advertiser to buy that space probably was digital, it's important to remember that "analogue" experiences, experiences we can't interact with in any way, still drive a lot of behaviour.

And that's the first important consideration: the world of marketing isn't just digital; it's not just about clicking on adverts. If your marketing division is totally focussed on digital technology, your organisation will be missing out on major opportunities. Instead, it is necessary for organisations to embrace communications that are both digital and analogue – and to create a consistency of approach across them both, with data collected from both that is then shared between them.

Talking to consumers

Your organisation will be communicating with its target audience in a number of ways including:

- **Relatively fixed assets**, like websites, directory entries and social media profiles
- **Online advertising assets**, such as banners and buttons
- **Fluid assets**, such as conversations with people on social media and through online chat and email exchanges.

All of these methods have their pros and cons. They all work in slightly different ways. Depending on your organisation, some channels will be more effective than others. It's a great deal to manage.

Digital marketing myth 2. We can measure how all our channels interact

Omni-channel marketing is a popular concept. It means using every single channel available, online and offline, delivering a seamless experience for the customer whether they are shopping in the High Street or online, and measuring the total effect.

This is a wonderful idea but, like many such concepts, in practice, it is very hard to achieve, at least at an individual level. If I see a TV ad for a pair of shoes, search for the price on various retailer websites, and then decide to go to a particular shop in the High Street to buy it, the maker of the shoes has no practical (and legal) way of tracking my journey. This is important, given that most shopping still takes place in physical stores.

Of course, if you have enough data about where and when your shoes were bought, how many people saw your TV advertisement, what people were saying about shoes on social media, the weather this week, the state of the economy and how many people visited your website, you can start to make connections. And technology can assist greatly with doing that. But the connections you make won't necessarily represent reality: you will still need humans to add common sense and experience.

Governance of marketing assets

Websites

Websites are of course essential as a communication tool for most organisations. So important, and with so many different potential roles (sales, marketing, recruitment, Corporate Social Responsibility, Public Relations, etc.) that, in some cases, numerous microsites spring up, each with a particular function. This can often happen as a way of bypassing IT Department processes that are, rightly or wrongly, seen as slow. Perhaps more often it happens because an enthusiast wants to create one for a particular purpose: DIY websites are simple to create.

Some large companies have literally hundreds of separate websites that have grown up over the last 20 years or so. It is impossible to manage so many, especially as in reality the existence of some may not be known to the people with responsibility for the digital estate. These *ad hoc* websites represent a potential risk in at least three ways:

- Some websites that have been built by enthusiastic amateurs *may not follow your brand guidelines*. Perhaps worse, they may be in contravention of the law, making false claims about your products or failing to be sufficiently accessible to people with disabilities.
- Some websites may be *leaking information by sharing data* about visitors with third parties. They will do this because they have third-party tags on them, little pieces of code that enable data to be collected and shared. Many websites will have a Google Analytics tag, for instance, enabling data on visitor behaviour to be accessed by anyone who has been given permission, including perhaps consultants who no longer provide services (I still have access rights to several websites owned by companies I haven't worked for recently). Often there are dozens of these tags and it's important to know why they are in place, who has access to the information, why and with whom they are likely to be sharing it.
- Some *websites may be insecure* and represent a way for hackers to enter your organisation or steal your organisation's data. We describe one example of this in the TalkTalk case study in Chapters 7 and 13.

Online advertising campaigns

Online advertisements are relatively cheap ways of displaying branding messages. And they seem to be easy to evaluate: if someone clicks on them, they have succeeded; and if they don't, they have failed. But they can also cause problems and so need careful management:

- **They may be placed next to inappropriate content**, as can happen when automated advertising placement tools are used; this type of automated placement is convenient in that it saves management time but the downside is potential damage to your brand.
- **They are very simple to create but easy to ignore**, compared with TV or press advertisements. And because online advertising campaigns so often focus on the visible returns of the campaign, there is frequently a focus on price over quality. As a result, online advertisements are often given to junior designers who may be skilled at tactical implementation ("if we do this, people are sure to click more") but lack strategic insight.
- **It can be hard to calculate their value**. Online adverts are often placed on parts of a page "below the fold" (necessitating the viewer to scroll down to see them); this can mean that they are not seen by a website visitor and so any payment for these might be wasted. Ads placed below the fold (see Glossary) might not get seen as much as those above the fold but they often have higher engagement rates presumably because users who bother to scroll are interested in the content of the page. Their value will depend on the nature of the payment: some online ads are bought on a cost per thousand impressions (CPM) basis so you will pay for them even if they are not seen; others are paid for every time someone clicks on them – cost per click (CPC). The latter might seem a safer way of buying advertising but, as always with digital technology, there's a problem: fraudsters frequently create websites to accept your advertising and then create robots to visit the adverts and click on them. This type of click fraud is already a multi-million-pound industry.
- **They can irritate people**. In an effort to "get the inventory away" they may be served to website visitors without impressions being capped at a certain number; this can mean that they are shown to the same individual numerous times, not just wasting money but irritating the viewer into the bargain.

Incidentally, if someone doesn't click on your online advertisement, that doesn't mean it has failed. A well-designed message will still affect the viewer's brand perceptions and may well prompt them to make a purchase when they next see the advertised item in a store.

One interesting tactic that is sometimes used in online campaigns is to offer consumers the opportunity to interact with the advertising. But consumers will have fun if they can. Let them manipulate your advertising and some of them are bound to find a way to embarrass you if you let them. Early social media experiments that displayed relevant social media posts in public places were quickly abandoned as insulting or obscene messages were submitted. Allowing the public their say in marketing is engaging and can be effective but it does demand proper oversight, an analysis of the potential risks and then policing the actual activity.

Social media

Social media accounts are almost as much of a requirement as websites. In fact, over 90% of companies in the USA use social media for marketing. The figure in the UK is lower with around 60% of businesses using social media (although the figure rises to 90% for the largest companies) (1). Given the very large number of consumers who use social media (around about one third of all people have a Facebook account) it certainly makes sense to consider social media as part of your marketing activities. But there are issues with it:

- **Not everyone uses social media**: While two thirds of British people do (the figure for Germany is less than half) that still leaves a third who do not. Using it for market research or monitoring consumer opinion may give you skewed results.
- **Those people that do use social media may not be open to a commercial message**: Perhaps they are socialising or gathering news and don't want to be bothered with advertisements. Just because you can reach people somewhere doesn't mean that it is sensible or cost-effective to do so.
- **Many companies feel that social media is less formal than, say, advertising on TV**: It isn't. But used carelessly, regulations get broken, data gets leaked, and consumers get irritated or feel patronised. Like any business activity, social media needs the involvement of experienced people who understand how it works and where the pitfalls lie. In other words, it needs proper governance.

But even if we accept that there may be problems, social media is a valuable resource for most organisations. At the very least, it is an effective way to enable consumers to get in touch if they have problems. There is a need for organisations to know where they have a social media presence and who controls it, because if the only person who has the password to the company Facebook account leaves suddenly, you may be facing some real difficulties.

There is however a particular issue with social media. Your employees use it. You are unlikely to find them creating websites or online advertisements that refer to your organisation in their spare time. But with social media, that may be exactly what they are doing. Their own personal social media accounts are likely to have places where they can upload personal profiles and these will often include the fact that they work for you and what they do for you.

Why is that a problem? Because they may also be using social media in a way that would embarrass you (and should embarrass them) – to tell offensive jokes or post offensive pictures, to bully people (including their colleagues), even to boast about what they are up to at work or, just as bad, even what they are feeling about work today (as if that was somehow, suddenly, not about you). In other words, social media is a place that personal and professional lives can merge or just collide and that's a problem. It is therefore essential that you have a written policy that details acceptable and unacceptable behaviour when using social media and what the penalties will be for a breach of this policy. This should cover things such as:

- Whether people can claim to work for your organisation in their social media profiles (probably OK on LinkedIn, distinctly risky on Facebook).
- Whether and how they can refer to their colleagues, suppliers, clients and industry in their posts.
- Who is permitted to post information on social media on behalf of your organisation.

Without a formal policy, that is then properly monitored, you may find that your colleagues say things on social media that your organisation has to take vicarious liability for.

Digital marketing myth 3. If you can't measure it, it doesn't matter

It's good to measure things in business because if you don't you can't manage them properly. And digital technology makes it easy to measure things, at least on the surface. You spend money on search marketing, people click on the ads, and you make a sale. If the value of the sale is more than the amount you spent on the advert you have made a profit! Easy. Social media is the same. You write some posts for Twitter, pay for them to get some coverage, people see them. You have strengthened your brand. Easy.

Except it's not so easy. Let's take search engine advertising. Sure, people saw the listing that you paid for on Google. But why did they click on it? Were those few words so enticing that people rushed to buy your product? Or did they search for your product and click on the ad because they saw a great TV commercial the evening before?

If all you measure is the journey from search ad to purchase, you are missing out a large part of the story. Of course, marketing departments generally aren't so naïve as to do this. They will attribute part of the sale to touchpoints they can measure, such as – social media activity, online advertising, previous visits to your website. They may even take some account of your TV advertising.

But when they do this, they are making assumptions about the proportion of value that each of those touchpoints delivered. Assumptions that can't be proved. They may be perfectly sensible assumptions. But they are not science. Some probing of such stats in reports may well be justified.

And how about social media? Those Twitter posts have been seen by millions of people according to the statistics. But what does that mean? Have millions of people actually read your posts? Or have those posts been delivered to a machine (a smartphone perhaps) where they remain unseen and ignored (although paid for of course)?

Additionally, it's as well to think about the value of what is being measured. A web page with lots of visitors is surely a better place to advertise than one with fewer. A PR agency with hundreds of thousands of followers is surely a better bet than one with followers numbered in the mere thousands. Not necessarily. It depends on the *quality* of the website visitors, of the social media followers. It is easy and cheap to buy website visitors just as it is to buy social media followers. The trouble is that those visitors and followers are likely to be students in the developing world earning a few dollars for clicking on web links and liking social media posts, or worse, simply computer robots. They are not worth anything to you as potential customers.

Of course, you may think that buying followers is a good way of making your social media profiles look more impressive, although you should be aware that it is generally fairly easy to spot that this has been done. There can be good reasons for doing this. But the danger is that this habit may lead to you becoming complacent about your social media activities.

SOCIAL MEDIA MATHS

How many people are seeing your social media posts? Let's do a little maths. We'll use Twitter as an example as with over 300 million monthly users it's an important platform used by a lot of marketers. But there's a problem:

- The average person spends 5 minutes a day on Twitter (some researchers say it is a good deal less than this).
- The average person posts on Twitter 5 times a day.
- The average person follows 200 people (again it's probably more than this).

If you are an average Twitter user you will be receiving 5 x 200 = 1,000 tweets a day. And you have 5 minutes (300 seconds) to read them, about 3 tweets per second.

There are two lessons here. First, don't believe social media "reach" figures, the exciting statistic you get when you are told how many people have seen your latest tweet. These are "opportunities to see" rather than a reflection of who has actually seen, let alone digested or acted upon, your post. And second, if you want your posts to have a chance of being seen (and there is little doubt that Twitter can be influential), you *have* to pay for them to be promoted.

Digital marketing techniques provide lots of data. But that's not the same as providing useful data or data that isn't misleading. In digital marketing, it's important to understand what the data really means. Media owners and marketing executives both have very understandable reasons to exaggerate the effect of their activities.

Reputation management

The internet has put a great deal of power in the hands of consumers. It gives them a platform to recommend products they like. And it allows them to take revenge on service that lets them down by posting negative reviews or critical comments on social media and even by creating websites that are critical of your organisation. (It's fairly easy for anyone to buy a web address like AcmeWidgetsAreRubbish.com and create a website that is highly critical of your organisation.)

And it's for that reason that every organisation needs to monitor what is said about them on social media (and the internet more generally), even if not using it proactively for marketing.

A single post from a disaffected customer is unlikely to cause any real damage, unless that person is someone famous and well respected with lots of people who follow them. And even if that post is shared a few dozen times it's unlikely that many people will see it. The danger comes when a few different people start posting about the same issue and the mainstream media pick up on this. When that happens the effect of a few posts can be massively amplified by coverage in TV programmes or newspapers that millions of people see. And that of course drives more people to the original posts and potentially to join in.

THE DANGERS OF UNFETTERED SOCIAL MEDIA POSTING

Ben Polis was the founder and CEO of the Australian energy broker Energy Watch. He had a "robust" sense of humour and posted his thoughts about the world including some very offensive comments on Facebook about Aborigines, Jews, Muslims, Asians, women, fat people and bogans – people that some others might consider to be unrefined.

Unfortunately for Mr Polis, the Australian media picked up on his comments. Outrage ensued and several sports organisations including the Melbourne Rebels rugby club cancelled their sponsorship contracts with his company. Worse, several energy companies severed relations with the company during the period of outrage, cutting Energy Watch's only source of income. (They resumed relations once things had quietened down – the new customers that Energy Watch provided were very valuable).

> Polis was forced to resign from the company he had founded. Was it really his fault? He denied racism and said that people had taken snippets from his Facebook posts out of context "and tried to say I was racist, sexist and misogynist, which is not true". What is true though is things that you say on social media have a horrible habit of hanging around and coming back to trouble you years later.

While it is important to avoid wasting too much time managing every negative comment (accepting that every negative comment still signals a place where you can improve your standards and/or messaging), when you feel that online activity is really damaging the reputation of your organisation, or individuals within it, then you might want to take some action.

Unfortunately, once something has been published online it is generally very difficult to get it taken down. You could of course take to the law and if you won a case then perhaps the offending comment might be deleted from the original place it was published. But the chances are that someone would have copied it first and shared it with lots of other people. Those unpleasant comments will be floating around the internet forever. So, not only will you have wasted your money, the mere act of going to the law to sue a disaffected customer for libel may well damage your reputation and get you seen as a bully.

What can you do then, beyond learning from the negative comment? The only real method is to work at posting lots of content that is favourable, or at least neutral, to your organisation. You do this in an attempt to make the negative comments become invisible by pushing them onto the second, third or fourth pages of search engines. Appearing in industry directories, creating company LinkedIn pages and Twitter profiles, writing guest blog posts and articles for popular media outlets in your organisation's name, all of these tactics will help you to create content that will appear nearer the top of the search engine listings.

Making the sale

The job of the salesperson is made immeasurably easier by digital technology. They can track down and contact many prospective purchasers using social media. And, when they have found them, they can potentially use machines (known as chatbots, see the section in Chapter 14 on automation) to do the hard work of selling.

Digital marketing myth 4. It's all about sales these days

What causes the inflated importance of advertising that is designed to generate sales, direct response advertising? Most marketing is designed to change people's behaviour or mindset in some way. Of course, it's good to have advertising where you can measure that change. And that's one of the reasons that certain digital marketing activities are so popular – they lead to a sale.

But while it may be harder to measure the effect of your TV ad than to measure the effect of your search ad, that doesn't mean you *can't* measure the effect of your TV ad. And it doesn't mean that your TV ad hasn't had an effect on people's preference for your products over a competitor's or on their likelihood to buy from you. Deciding to stop using marketing channels that don't lead directly to a sale can be a huge, strategic, mistake.

Sales technology

Analytics

Data analytics provides an organisation with real-time information on its performance. With more information available, sales managers can start to see patterns that could be exploited. Different techniques can be compared, and the better ones adopted. Information can be shared across the team stimulating not only team spirit but also stimulating innovation as people are exposed to new ideas and situations.

And data makes it easier to recognise previous customers and view data about them – what they bought, when they bought it, perhaps even what stimulated them to buy. This makes it easier to craft effective sales messages. And the selling process is made even easier if you can demonstrate to your prospect that you know something about them, especially if they have been a customer before. According to Accenture, a majority of customers are more likely to buy from a shop that knows their name (56%), offer recommendations based on a previous purchase (58%) or simply know their purchase history (65%).

Improving the buying experience

Another thing that increases sales is a positive and easy user experience and digital technology can help here too. Many McDonald's fast-food restaurants have interactive screens where customers can order and pay, saving themselves time being jostled in a queue. Of course, some people prefer to talk to a human who can answer questions and take orders so the company employs these people too; but it needs to employ fewer of them, thus saving money at the same time as increasing overall customer satisfaction.

Digital technology can also help you personalise the buying experience, enabling you to see how you might look with the item you are considering. Clothes, glasses, make-up – it's common for people to model these items online. But the same technique – manipulating an image of themselves that the prospective customer has uploaded can be used for many other things – cars (don't I look cool), holidays (I must be enjoying myself), films (wow, I look scared), almost anything at all.

Pricing

Manipulating pricing can be a powerful selling tool and many organisations take full advantage of this online. This can be as simple as creating algorithms that change prices depending on availability. This type of dynamic pricing is frequently seen in the travel industry where typically prices are low a long way before the date of travel to attract people in, rise as the date of travel approaches and seats fill up, and then falls again just before the date of travel so that any empty seats can be sold.

But dynamic pricing doesn't just depend on timing. It can depend on what you have bought before from a website or whether you have looked at an item on their website recently: if the site thinks you are keen on something you may have to pay a higher price than someone else. It can even depend on location with prices varying between countries or regions. (There is nothing odd about this – insurance prices are frequently dependent on where you live.) To be successful, dynamic pricing depends on sophisticated data collection and analysis.

If you are changing prices frequently you need to ensure that they are displayed efficiently. Easy to do on a website, less easy in a store perhaps.

But in-store digital displays allow High Street retailers to adopt this tactic. The American supermarket chain Kroger, for instance, has digital shelving which can display prices together with product information. This enables the stores to change prices efficiently and dynamically, depending on demand. There is even a potential for these smart shelves to send advertisements and personalised special offers to the mobile phones of individual shoppers.

Digital marketing myth 5. We need to build relationships with our customers

People don't trust advertising anymore. Well, that's true to a degree. Trust in advertising has fallen to all-time lows. And while people appreciate it for the information it delivers about new products, and even (in the UK at least) for its entertainment value, they are increasingly fed up with being bombarded with advertising, especially those "creepy advertisements that follow you around".

As a result, some people would have you believe that no one ever responds to advertisements (this is patently untrue) and that the only way forward is relationship marketing.

And it is worth it for some companies to build a good relationship with their customers: lawyers, banks and car manufacturers, for instance. But most companies don't need to do this. That's simply because most of their consumers don't want a relationship with the brands they buy. (I really don't want a relationship with my favourite brand of baked beans, for instance, and I am probably not alone).

Nonetheless, the "build a relationship" advocates have many followers. Many companies feel that the best place to build relationships is social media. They therefore invest a great deal of money sharing stories on Instagram and Facebook and building a personality that is almost human. And some would say this is an effective strategy. But there are a number of companies who have found that this type of activity is merely a distraction from the important business of building strong brand values online and offline.

Digital technology makes it superficially easy to communicate on a one to one by email or social media or even by behaviourally targeted advertising (see Glossary), but that's not the same as building a relationship.

Finding the right people

Building talented teams is always a priority for managers. This used to be relatively simple in marketing. But digital technology has widened the skill sets needed. As well as understanding basic marketing concepts such as the brand and consumer demand, and alongside skills such as communication, analytical thinking and creative problem solving, competent marketers need to be able to understand a wide range of technologies including how to set up, track and evaluate online advertising, email campaigns, social media activities, and search engine activity.

Each of these will have different processes: different commonly used tools, different ways of evaluating success and failure. And each of them can be used to generate sales or strengthen brands. Digital marketing is immensely complex. And it sits alongside non-digital marketing activities (TV, posters, press, direct mail) which aren't going away any time soon but which are increasingly impacted by their own forms of digitisation.

This increased requirement for technical competence among marketers is driving increased specialisation: people are no longer markets, they are search engine specialists, social media and content marketers, email and data scientists. And because there is a lot to learn about each individual discipline, it is increasingly difficult to get people to understand how their piece of marketing, important though it is, works alongside (and, more importantly, in combination with) other types of marketing and how it fits into the wider marketing goals.

Senior marketers can often have blind spots as a result. One area that has been problematic in the past is social media. In a way, this is a truly new form of advertising. Online campaigns equate to traditional press campaigns; email campaigns equate to direct mail campaigns; even search can be compared to classified advertising. Marketing directors understand these things. But how does social media fit in?

The temptation is to say "the intern must know how it works". And perhaps it's true that the most junior, and typically therefore the youngest, staff are the most familiar with actively using social media. But they are also inevitably less familiar with marketing disciplines. In one technology company that I worked for, I discovered that the junior executive responsible for social media had started writing posts about the singer Gloria Estefan. When I asked why, the response was that she was trending on social media.

Unfortunately, there had been no attempt to link the technology company or its products with Ms Estefan. This activity, based purely on what internet users (not just customers) were doing at a particular moment, had absolutely no marketing value. Was this the fault of the junior executive, or the fault of the person who had given them the social media task without putting appropriate structures and quality controls in place?

Getting people who can use digital technology effectively is a constant difficulty for most organisations. So much so that it is sometimes worth acquiring another organisation simply for the digital talent it contains. We will explore this, and other technology issues around mergers and acquisitions, in the next chapter.

Reference

1. According to Turnerlittle as described in: Hebblethwaite, C. (2019). *Only 60% of UK Businesses Using Social Media*. [online] Marketing Tech News. Available at: www.marketingtechnews.net/news/2018/jan/04/only-60-uk-businesses-using-social-media/ [Accessed 24 May 2019].

7

THINKING DIGITAL IN MERGERS, ACQUISITIONS AND VENTURING

Summary

Mergers and acquisitions happen for a variety of reasons but whatever the driver there is always a need to ensure that business systems, including digital technology, can be integrated efficiently as part of joining two organisations. And this will be the case even when the reason for the merger is the acquisition of new technology used by or owned by the target organisation.

Getting to an agreement that will work from a technology perspective is complex. Once analysis has been made of the potential technology fit of the target organisation, there will be a need to consider how technology integration can be implemented. Care will need to be taken if the merger is happening to acquire new technology or technology skills as too little flexibility here, e.g. imposing the dominant company's systems may well kill off the benefits that would accrue from learning about new technology. A data room will be needed to record any technology issues and an audit of the digital estate should be undertaken. The final deal should specify technology governance issues. After the agreement is signed there is work to be done aligning technology cultures and strategies and integrating systems.

We discussed in Chapter 2 the attributes that make up an effective framework for delivering digital governance and in Chapter 10 we will describe the means to ensure that your organisation stays compliant with the expectations put upon it by you, your partners, the government and the public at large.

This chapter examines how digital technology and its use is both an obstacle to mergers and acquisitions (e.g. through different legacy systems and thus different processes) as well as, when properly managed and governed, a considerable asset both to the merger or acquisition process itself and potentially to the new organisation that results from it.

There can be a number of reasons for merging or acquiring, including:

- **Synergy:** The merger will increase efficiency, decrease costs and have a stronger position in the market.
- **Diversification:** The merger will spread risk into other areas.
- **Growth:** The merger will make the organisation bigger and give it more power in the market (e.g. with suppliers) and/or access to new markets.
- **Eliminating competition:** The merger will kill off a competitor before it gets troublesome.

In our digital world though, we can add another probable benefit:

- **Obtaining advanced technology:** The merger will access cutting edge tooling and the teams that constructed it.

Indeed, Accenture Research in 2018 (1) established that digital motivations are increasingly driving merger and acquisition activity. Forty-eight per cent of organisations surveyed stated that a major goal was to gain next-generation technologies whilst 39% acquired new (technical) capabilities.

For example, in 2017 Amazon bought Whole Foods. This was a great deal for both parties. Whole Foods got an immediate online presence from Amazon and (no doubt) highly cost-effective hosting of their IT services on Amazon Web Services. In return, Amazon obtained a whole new market segment, a new cross-selling opportunity as well as a better appreciation of bricks and mortar retail. The latter was itself vastly important: Amazon will, after all, be competing with the High Street for decades as traditional shopping (including click and collect) still accounts for over four fifths of retail activity in the UK.

Acquiring services rather than organisations

We will start this chapter by considering what it means to acquire (or outsource) a service. Today's world is increasingly typified by outsourcing, co-sourcing, partnership, collaboration and other co-operating models, all of which are made easier through digital technology. All these have attractions, but they should be considered carefully, as they represent one degree or another of you losing overall control.

To one extent or another, outsourcing deals are still amalgamations that link your digital operations with those of another organisation. Recognising them as such, and consequently applying the suggestions in this chapter to the initiative, will significantly help.

The only difference between an outsourcing deal and a straight merger or acquisition is that you are likely to have, at least during the tendering stage, multiple 3rd parties to negotiate with. Your board is still ultimately accountable, via your identified owner of the initiative. However, it will be best if you have one appropriate manager as the lead for each negotiating party reporting to an Accountable Officer.

Assessing the fit

Merger and acquisition activity has, for too long, traditionally been rather dominated by finance and business managers. Digital technology changes the dynamic significantly.

Whether you are acquiring or being acquired, this activity is still principally about the capture of knowledge, of information and of data – whether that is about products, services, clients, techniques or patents. The intellectual property involved in a merger has an intrinsic value that you will want to protect and maximise. So too will the skills in using the technologies deployed.

Therefore, when assessing a potential merger, you need to consider the issues that relate to the fit of your organisation with that other organisation.

Issues that will have an impact on your digital governance and digital technology plans will include:

- Does the target organisation own technology or patents that should be kept away from competitors?
- Has the target organisation gained useful experience of, and skills in using, its digital technology that would benefit your organisation?
- Does the target organisation own technology assets that you could use or extend, such as data or services that it has created through the use of technology?
- How aligned, or not, are the two organisations' views and practices in regards to such matters as privacy, ethics, culture, security, resilience and compliance?

Making it happen

Therefore, during this most critical of times for an organisation, you need to charge your top management team with extra responsibilities. There's certainly no more important time for assuring the effectiveness of your digital governance. Most often, you will want to adopt a staged process that is measured as the merger can move between stages. This means that you can have confidence that the costs and benefits, including those related to digital technology, remain within the parameters that the governing body originally established.

Maintaining a focus on digital

You should obviously give proper consideration to the challenges of any merger. But there are some major issues of digital practicality around merging two different digital set-ups.

Let's think about customer databases as an example. Your organisation has one, and so does the organisation you are acquiring. You are probably anticipating the need to merge and rationalise them. Unfortunately, that could be convoluted and very complex. Perhaps the customer names are organised differently or perhaps the two databases use different software.

But is it really necessary to go to all that trouble? Database technology is far more sophisticated than it was and there may be tools that can provide effective searching or dashboards across both databases without having to merge them at all. If that's the case, a better solution (and one that would also provide an additional measure of resilience) might be to keep the two databases separate, adding new customers to one database and merging records into that database, with the other eventually being turned off.

Another major practical challenge is "shadow IT". The use of locally grown, operated and maintained systems often increases in the immediate aftermath of mergers and acquisitions when people are a bit cynical of what is happening and try to hold on to the things they know and like. This is especially so when new digital technology is forced upon them by the dominant partner in a merger.

Shadow IT can destroy some of the operational efficiencies a merger is designed to deliver. In order to reduce it, you need to ensure that the benefits and objectives of any new systems are fully understood and accepted by the people who will be obliged to use them. You need to ensure that a commensurate level of training and local support, e.g. through system champions, is made available and taken up.

The adequacy of your resources for this critical activity should be a prime concern for your colleagues on the board. You may or may not have personnel with the requisite skills in your organisation. But have you enough of them? Have those with the requisite skills shared them sufficiently across the team? Is everyone working to the same agenda and using the same scoring criteria, etc.? It is always best to be definitive in these areas.

Corporate venturing and start-ups

Of course, you don't have to buy a company outright and you don't always need to be looking for a financial upside. Venturing can be about accessing new digital technologies and new business models in a low-risk way. This is what, for example, the aerospace company Marshall of Cambridge is doing with its investment vehicle Martlet. Investment in a strawberry picking robot might not seem obvious but it could surface new ways to process and handle delicate aircraft parts.

Even if, as the venturing party, you decide to take a largely hands-off approach to benefitting from their creativity (perhaps providing mentoring, premises, financial services, and administrative help, but not strategic management), applying good digital governance over your interactions will still be appropriate. Indeed, digital governance advice itself is likely to be highly beneficial to both your partner and to the valuation of your joint venture. At the same time, too active an involvement in a venture and over-rigid operations models may stifle the very innovation you seek.

Certainly, the traditional approach of valuing a company on the basis of future cash flows or earnings is becoming harder and harder to sustain in the digital economy. In fact, in 2017, over three quarters of companies that sought stock market listings in the USA were unprofitable. (Compare that with the situation 30 years earlier when the reverse was true.) As digital technology becomes more pervasive, and emerging companies are based on ideas and projects, new valuation models for such operations will be needed, e.g. around best-case scenario payoffs. Such payoffs can only be maximised when digital governance has been a key feature of operations.

But of course, if your aim is to benefit from the creativity of a start-up, or to understand the technology it is using, then any final valuation may be of little importance as the value of your investment is not in the financial return. If that is the case, then you need to accept this and regard your (probably small) seed capital investments as ways of learning rather than financial investments. That should affect the governance approach you take to this sort of investment.

Data disclosure and data rooms

An increasingly common approach to facilitating merger, acquisition and venturing discussions is the creation of a data room containing all of the information needed to facilitate the discussions needed to deliver a mutually satisfactory deal. Once these would have been hard copies in an actual room but are mostly now virtual documents in a virtual room, introducing a new digital dependency and one that is of critical importance.

Working together, the governing body and top management should identify what information they need to disclose to the bidding parties. This applies equally to bidders, who should identify what they expect to be given. Certainly, before anything else is done, bidders should undertake an evaluation of the data room contents so that early requests can be placed for missing information. The nature and extent of such gaps will, of course, be an immediate indicator to you of potential issues with the organisation's digital governance.

You may not want your wider organisation to be involved in, or aware of, the merger and acquisition discussions. In addition, the concentrated nature of the data in the data room will make it an unusually attractive asset for those with the urge to obtain it (rival bidders perhaps, disaffected employees or simply criminals).

The ten commandments for data rooms

Here are ten (digital governance) commandments that you would be well-advised to follow in establishing and operating a data room:

1. **Naming your data room and deciding where to host it**. Naming your data room something like "Acme Widgets buy out of WidgetsRUs – Data Room" is clearly not terribly clever in that it advertises the presence of data that is probably very confidential. It happens more often than you'd imagine. You will need to store this data some-where, possibly on your own servers, with a third party or in the cloud. Wherever you store it, ensure that it is secure from accidental discovery or deliberate theft.

2. **Service availability hours and arrangements for continuity**. Do you need 24x7 availability and if so why? How will you ensure that you can recover access to the data room easily after any technology fail-ure? You need the data in the room to be backed up on a regular fre-quency if there are technical failures so you will need to decide what the frequency will be.

3. **Partitioning the data room**. If you have multiple bidders, you will have multiple sets of the data in a number of rooms. You will need to secure access to the individual rooms if Bidder A is not to see what Bidder B is saying and doing.

4. **Information classification, marking and handling**. The general importance of this process was described in Chapter 2 but it is even truer here. Simply marking everything "confidential" and having a non-disclosure agreement in place means little or nothing. You may need several marks as some data may be intended for public use, some for internal use only and some for you to share with the merger target. You will need to be sure that all parties understand and agree on what handling a given marking allows and does not

allow. And you will need to be confident that all parties will deliver those behaviours.

5. **Content management**. You will need to carefully identify and validate what should be in the data room, and what is in it. But when you need to update a file, you need to know how this will be done and by whom. What document version controls are in place? And how will both parties be satisfied that the integrity of all the files has been retained?

6. **Managing inclusions, removals and updates**. As well as the specifics of content management, who governs the process? Who approves the changes that are made to data room contents?

7. **User roles and access controls**. Not everyone has the same roles, focuses and responsibilities. Two people with the same job title on either side of the bidding process may have different roles and therefore different access rights. These roles need to be identified and approved and access controlled, with the results being subject to top management scrutiny and approval.

8. **Remote access security and logging of activity**. Both parties will need secure access to the data room. At least one part will be accessing the data remotely (both if the data is held in the cloud) and of course this access needs to be secure. What approved users actually do, the specific actions they take within their approved roles, should be logged so that you can track what has happened if there are any disputes.

9. **User deposits**. As well as the management of the process of file inclusions, etc. (see 6 above), arrangement needs to be achieved within the parties about what data room user roles can actually add approved files. Even if the results of analyses of data room content are stored on the bidder's own technology infrastructure, users will need to know that is what you require to have happen and where.

10. **Data room close down**. With the deal done, it is tempting to just turn off the data room. However, there may be a reason to return to the data in the future, e.g. to resolve disputes. You will therefore need to make a decision about whether, and where, to archive some or all of the data. An archive of the data might still be discovered by other unauthorised people, as indeed may any backups. In particular, if you have used a third-party cloud server to host the data room, backups

may have travelled into distant geographies if you have not prohibited that. Certainly, you should have a plan for the ultimate removal and destruction of data room contents.

A good way to orchestrate all of the above would be for you to create and publish a *Data Room Policy*. Like all other policies, this should be approved by the board as a key-stone of its digital governance.

Due diligence

Conducting due diligence on the digital estate (of the target organisation's databases, communications systems, operational software, processing platforms, data assets, etc.) may well be painstaking work especially if the organisation has poor digital governance! It cannot be rushed, however much it is tempting to jump into the face-to-face meetings and commercial discussions that are dependent upon it, as it is going to be increasingly fundamental to the overall result.

You should therefore ensure that the digital due diligence work is:

- **Fully planned** with a clear process being followed, and with appropriate decision points about valuing technology as part of the potential deal. For example, how and when are you going to value the customer database your target owns – is it comprehensive, accurate, legal to use?
- **Sufficiently rigorous and comprehensive** with everything covered. For instance, are you sure you are aware of all the websites that the target owns, including the unofficial ones and the obsolete ones?
- **Sufficiently resourced** and with appropriately skilled resources being used. For example, does your organisation understand how the software you may be buying should be maintained and whether that introduces any dependencies on 3rd parties that would then need to be researched and subsequently managed?
- **Specifically managed**. Is the person accountable for this merger, acquisition or venture capable of taking account of digital issues even if they are not a specialist themselves?
- **Formally reviewed**. Is there sufficient technical challenge and validation of results or are those complicated and boring (in the eyes of some) technology issues nodded through without sufficient consideration?

- **Formally approved**. Have those results both been adopted but also reflected into other aspects of the deal?
- **Translated into risks**. Have digital risks, including data breach risks but also digital dependency risks, been adequately described, weighted and mapped into value exposures consistent with the contractual values?
- **Clearly documented** and tracked in the subsequent decision making and negotiations.

In 2016, the British telecoms company TalkTalk experienced a sustained services failure, in the form of a major data breach (see the text box on page 228). It was subsequently identified that the outage had been caused by the injection into the system of malware (malicious software). This happened because a server had not been patched with an appropriate software update. This server was a legacy from Tiscali, an ISP (internet service provider) company that TalkTalk had taken over. Unfortunately, the server's existence hadn't been flagged up and it wasn't included in the IT support contract.

Deal provisions

The eventual contract that comes about to enact the merger or acquisition must address the full range of digital governance issues that we have set out in this book. You should ensure that there are unequivocal contract schedules or other documentation identifying:

- The digital assets involved
- The digital services being delivered, and the service levels agreed
- The security and privacy requirements of the information held
- The arrangements to deliver digital resilience
- The processes for managing future problems concerning digital technology including the financial exposure to each party caused, for instance, by penalties for data breaches, by technical failures or by unexpected maintenance costs.

It will be important to ensure too that there is consistency between the criteria to be used to evaluate, manage and subsequently pay penalties for the occurrence of digital and non-digital incidents. There should similarly be consistency across different types of digital incidents, whether these relate to operational events or security, privacy or resilience events. After all, you will not want to have an incident costing you more simply because someone chose to classify it as a security incident rather than an operational incident when the end impact on the organisation will have been the same.

The deal's contract provisions also ought to make clear who carries the financial exposures for any failures or limitations in due diligence – especially if those were imposed by the contract awarding party.

MANAGING UNPLANNED COSTS

In the late 1990s, Siemens effectively acquired BBC Technology, the technology arm of the BBC, the UK's national broadcaster. At the time, there were some material limitations as to what could be achieved in even the most thorough due diligence. This was made more complex by the fact that "the show has to go on" even when there might be technical problems.

The ultimate solution adopted was to contractually agree on a discovery period during which costs identified by type but otherwise unknown were still attributable to the BBC. But this period had a sunset date after which Siemens carried any unplanned costs that were then identified.

The deal is done, so now what?

Post-contract evaluation

Due diligence is always conducted in a time-pressured environment. No doubt agreeing on the acquisition price, whilst based in large part on the due diligence findings, requires you to make a judgement call drawing heavily on your experience and overall confidence level. However, the changing nature of digital technology serves to complicate matters. The value of any digital assets you are buying may well have dropped considerably since the start of negotiations. In addition, the points of difference that your acquisition originally boasted may have been rapidly undermined by new services and products delivered by new digital competitors.

Cultural evolution

The success of mergers and acquisitions ultimately comes from the people that create and use the organisation's information and intellectual property (IP), as well as the systems that are used to do so. And, of course, it is about two sets of such people. You will have different cultures to combine, with different heritages of technology experience and different technology skill sets.

Whether or not acquiring new systems and better ways of working are part of the justification for the new organisation, there will always be systems for you to bridge, merge or replace. And people have to use those systems. One party's staff may well feel fear of the unknown and upset about what will probably be imposed upon them – perhaps because they don't trust it, perhaps because they feel it will be difficult for them to learn the new skills required. Making sure that people who are coming new to digital systems, or new digital systems, are properly trained in *why* they are useful as well as *how* they are used, is essential if you are to get their buy-in.

Some of the behaviours you will have inherited will potentially be negative, reflecting entrenched views of a true past, of old poor digital systems, which need not be true of the future. At the same time, or alternatively, however, this might be true of your existing staff too. There will be no time like the present for you to formally evaluate and identify the positive aspects of both, and their similarities, so as to maximise them, whilst you also work to address the negatives and evolve the differences. This will need a specific plan and is likely to require specific funding – funding that you should have allowed for in your acquisition budget of course.

INTEGRATING ACQUISITIONS

In 2008, BAE Systems, formerly British Aerospace, the global aerospace giant, acquired Detica, a major information technology, cyber security and fraud detection company. At the time, Ian King, Chief Operating Officer of BAE, was reported as saying that the acquisition would help it expand its security business, particularly in the US Homeland Security market.

But this was an initiative that was trying to merge a traditionally run engineering conglomerate with an essentially UK-based, and relatively small, niche consulting business. This was always going to take time and considerable effort, even once this reality was even recognised.

> The acquired business now indeed trades very successfully as BAE Systems Applied Intelligence, but it did take some 2 or 3 years for BAE to employ Detica resources, and its state-of-the-art digital technologies and skills, for the internal tasks in which Detica was expert.

Technology strategy

Your strategic plan that envisaged this merger or acquisition may have anticipated many of the issues we have outlined. You hopefully also recognised that the world after your deal was concluded would actually turn out to be somewhat different from the one you envisaged! The effects of technology, even in the short term, are hard to predict. You should specifically address these matters in an update to your technology strategy.

For example, deciding what locations you will now operate from, and what processes you will employ, is undoubtedly affected by many considerations. But effective digital governance means ensuring that your decision making around this gives meaningful, and increasingly a high priority to, consideration of the digital issues. In this instance, it could give you real gains in resilience.

In some ways this will be simple. For instance, there may well be other easy ways of enhancing your organisation's digital resilience (see also Chapter 14) at a lower cost than otherwise might have been the case. You might also be able to retain some parallel systems, some duplicate processing platforms and perhaps even some data centres.

Other areas may be more complex. For instance, you will need to evaluate whether your technology strategy or the acquired organisation's is the stronger. Just because you are the dominant party does not mean your technology solutions should be preferred. The other party's may be a better technical solution. However, if your operations are larger, and you have many more people, there will be a trade-off between getting better results from using better technology and losing money because you are having to train large numbers of people in new systems.

Whichever set of technology wins, there is no doubt that merging two environments together can often be extremely expensive and may well deliver mixed results.

TSB BANKING FAILURE

In September 2018, customers of the UK challenger bank TSB experienced problems with online and mobile banking. In an earlier incident in April 2018, 1.9 million TSB customers were locked out of their accounts. These incidents cost TSB at least £175 million, lost them 6,000 customers and obliged them to hire hundreds more customer complaints staff.

The incidents have been traced back to the migration of customer data from previous Lloyds Bank systems to new systems owned and operated by Sabadell, the parent of Santander and owners of TSB. These incidents are the latest events made public in a 5-year saga of integrating two banks' systems.

As ever, it is always a case of measured decision making, avoiding the trap of just falling for the obvious, and then managing the chosen process extremely carefully, with full engagement by the board in the consideration of the digital governance considerations.

Post-implementation review

Once the deal is in place, it is all too common to be tempted to "just put your head down and get on with it". However, a post-implementation review is good practice. And when this happens you should ensure that the digital aspects of the deal get as good an airing as the financial aspects. In order for this to happen you will need to ensure that the review team has appropriate digital skills within it.

Once the review team reports you will then want to ensure that there are sufficient digitally skilled resources available to address the team's findings, some of which will also inevitably point to areas on which digital governance can be strengthened.

Finance departments will be intimately involved in mergers and acquisitions, and in the technology that can drive them. Of course, in many cases finance departments are highly skilled with digital technology and have been responsible for taking the first digital steps in their organisations.

But the use of digital technology in finance departments doesn't end at accounting software, as we will see in the next chapter.

Reference

1. Accenture. (2019). *Digital Consumer Survey 2019*. [online] Available at: www.accenture.com/us-en/insights/high-tech/reshape-relevance [Accessed 24 May 2019].

8

DIGITAL TECHNOLOGY IN ACCOUNTING AND FINANCIAL MANAGEMENT

Summary

Finance departments have often led the digitisation of organisations with the automation of accounting systems. But there are many more opportunities for the CFO to grasp opportunities that can be strategically significant.

One opportunity is to collect and use the increasingly large amount of data that organisations have available. Much of this data can be used to build better, more complete, financial models that will help organisations deliver more effectively. With more data, more up-to-date data and the ability to analyse data on the fly, CFOs can detect significant new trends and support making better decisions. System automation is another opportunity which not only reduces costs but also increases the accuracy of data and makes it easier to compare different data sets. However, making the most of these opportunities requires planning as well as common sense, together with a degree of scepticism about the accuracy of computer-generated models.

Finance departments have led the way with digital technology. Adding machines and punch-card systems were used by accountants in the late 19th century but the first recognisable computer that was used for accounting was built in 1955 to process the payroll for a General Electric factory. This used magnetic tape and could process the factory payroll in 40 hours. We have come a long way since then. Spreadsheet software (Visicalc) was launched in 1978, which allowed some financial modelling. And, in the same year, Peachtree launched a package of accounting software that would run on early personal computers – avoiding the cost of buying an expensive mainframe computer. By the mid-1980s, PCs were found in millions of finance offices around the world.

As a result, finance professionals, perhaps more than any other members of staff, are comfortable with computer technology. And perhaps that's not always been such a good thing. It has meant that early IT departments were often managed by Finance, but thus with a focus on cost control rather than strategic development. Indeed, it may have led to some complacency in the finance community itself, which now often results in finance professionals lagging behind their colleagues in operations and in marketing when it comes to the use of digital technology.

But that's changing. Driven by a desire for greater efficiency, speed and accuracy, finance departments are increasingly adopting new digital tools to improve the service they offer to the rest of the business. And the role of the Chief Financial Officer (CFO) is changing – from making decisions based on data from earlier periods to using predicted future trends to guide the business.

The problem

The finance department is at the heart of any organisation. But a great deal of the work they do, especially around accounting, involves intensive manual transaction processing. This takes time, is backwards-looking and, because it involves humans, it can be inaccurate or inconsistent.

In addition, traditional financial data, related to hard things like money, assets or production units, can only ever give a limited picture of an organisation. For instance, the sentiment expressed on social media can affect share price, sales volumes or services take-up. But you will rarely find this analysed by the CFO.

> **ONE TWEET KNOCKS $100 BILLION OF THE NYSE**
>
> Back in 2013, a single tweet knocked over 1% off the value of the New York Stock Exchange. The Twitter account of the much-trusted Associated Press news agency was hacked by a group calling itself the Syrian Electronic Army. They tweeted that the US White House had been bombed and that President Obama had been injured. Panicking investors sold heavily for several minutes until people realised that the news was fake and after which the market recovered.
>
> A similar, and only slightly smaller, event happened to Twitter itself in 2015. Disappointing results were posted accidentally and for only 45 seconds on a Nasdaq-run investor relations web page about Twitter. At first, no one seemed to notice. But a single tweet from Selerity, a document scanning service aimed at investors, publicised the disappointing results at the time when the markets were still open but Twitter had yet to explain its results. At one point, shares dipped by 25%, wiping $8 billion off Twitter's value.

In fact, in any organisation, there are likely to be multiple sources of data that are not analysed by the finance department, but which could have a major effect on success, and therefore which should play a part in financial decision making.

Another problem is that in larger organisations, different business units, or different subsidiary companies, may have different processes, sometimes caused by different management reporting requirements. This can obscure the truth and cause difficulties for financial decision makers.

And then there is the nature of financial data. Generally, it is historical, based on what has happened in the past. Of course, projections from this data can be made. But even if people trust them, projections take time to create and the data used to inform them may be obsolete by the time the projection is used by decision makers.

Better decision making by the board

Digital technology enables better financial decision making by the board in a number of ways:

- The automation of risk assessments and the automatic signalling of worrying changes; this can be combined with the inclusion of tolerances and traffic light reporting
- The use of dashboards and data visualisation to make data easier to understand
- Real-time financial data, rather than periodic reporting, so that boards always have the latest data in hand
- The ability to model different assumptions immediately, e.g. financial models can be rerun to reflect changing exchange rates or tax rates
- The ability of the board easily to zoom in and out of the organisation, for instance to compare individual operating units
- The ability of different operating units to self-serve their financial activities, rather than asking HQ finance departments for advice; financial feedback loops can be created where financial analysis and predictions are delivered back automatically to operating companies
- The ability to analyse how people are responding to your public financial reports, by analysing the sentiment of news articles and social media
- The ability to make inferences from the language used, e.g. by competitors in their public financial reports and press releases.

The opportunity

Digital technology brings an opportunity to make financial processes more efficient – more accurate, more consistent, more cost-effective. But it also brings an opportunity to make those processes more flexible, with rapid (sometimes real time) processing and the ability to include or exclude different sets of data or information in models and predictions at the click of a mouse.

As with many other parts of an organisation, in finance the eternal battle between efficiency and agility can, at least to a degree, be solved by digital technology. But, in addition, technology can directly help CFOs and their teams. How is this? One way of looking at CFOs is to allocate four different roles to them:

- **They lead**. They are strategists who help the board find the best way forward for their organisation. Digital technology enables them, through powerful data analysis, to provide detailed insights, accurate predictions and justifiable recommendations that would not have been available 20 years ago.

- **They enable**. CFOs enable company operations to happen by allocating resources: for instance, they provide money for technology and communication systems. And though data analysis and the development of computerised projections they can do this more effectively, deciding where resources should be allocated and predicting the likely effect on the business of investments.
- **They protect**. The CFO is the steward who preserves the financial assets of the organisation. They can (or should) use digital technology, including cyber security technology and data analysis that looks for unexpected events or inconsistencies, to prevent fraud.
- **They do**. The CFO's team are operators who undertake and record transactions. In many cases these operations can be made more efficient through digital automation.

There are a number of digital technologies that can assist finance departments. The most important can be grouped into two related areas:

- **Data and information:** The collecting and analysis of data, even if it is unstructured, and the creation of insight and forecasts based on this data. This is especially so in a world of Big Data.
- **Automation:** The use of robotic automation for some processing tasks, often assisted by artificial intelligence.

Data and data analysis

Data and information are not just used to provide audit paper trails. They can be used to deliver market insight and business forecasts that can impact on decision making. On its own, data is fairly useless. But combined with data analysis techniques enormous value can be found.

Data analysis gives you the truth. As organisations get more complex, with data stored in different places and different formats, it can be difficult to discern what is the truth. Is a profit being made on manufacturing the new widgets across all of your subsidiaries? Is the WidgetsAreUs subsidiary in Derby really so much more profitable than AcmeWidgets in Pittsburgh?

Was the drop in widget sales in Spain this summer caused by the weather or were there a combination of factors that produced the problem?

Technology helps you avoid data silos and analyse unstructured "data lakes". Sophisticated analysis of massive sets of data, including non-financial data such as social media or unstructured data like HR records, can provide a more accurate version of the truth, helping you to understand where value is really coming from. New types of analysis such as the sentiment analysis of online text can identify market changes and even point out unethical behaviour. And by ensuring that data is analysed in the same way across the organisation (to thereby find variances) you can create a single source of truth, something especially important in conglomerates and international organisations.

Data analysis helps you to identify trends from past data, and thus predict the future more reliably. There are obvious benefits here in financial risk management such as forecasting which of your customers might not pay you. Forecasting can also help you manage investments. We live in an age where new breakthrough ideas that look good are valued highly even if they don't have a certain financial future. This is because ideas that might disrupt a market can give your organisation massive increases in value (even if they probably won't). Everyone wants to be the next Amazon/Google/Uber. As a result, less innovative, but also less risky, ideas can get ignored. But better data analysis will enable you to model the value of everyday innovation and prove its worth.

Data analysis gives you flexibility. With improved analysis you have the ability to develop and analyse different scenarios. This can help you to manage risks, both by predicting the outcomes of decisions and also by prompting you to practise responses to situations that may not have been anticipated until you ran your advanced analytics. Combined with machine learning, data analysis can help you better understand the financial impact of operational decisions.

Data analysis gives you speed. Faster processing of information means that you can confer with key stakeholders more rapidly and make faster decisions, giving yourself competitive advantage. The ability to extract data during a process (such as during the manufacturing of goods), rather than after the process has finished, adds to this advantage. For example, production machinery may be linked to the internet in order to allow engineers to predict mechanical problems, and this same data, extracted as the machines are working, can also be used in real-time financial models.

Data analysis gives you warnings. If you can detect anomalies – unusual behaviour by individuals or within organisational results – you might be prompted to investigate things that you might not otherwise have noticed. You might even be able to put a hold on certain actions (such as an unusual payment to a payee overseas) until an investigation has been conducted. This can help you to detect and prevent fraud as well as major market changes that could be about to cause problems.

Financial data pre-requisites

There are some requirements though, if data analysis is going to be effective. Any financial data presented to the board must be high-quality data. You need to know, or to be able to ask, whether the data is complete, accurate, up to date and unadulterated. No market data set can ever represent the truth completely, but you need to be able to judge its quality.

You also need the right analytical tools. And to go with those tools you will need the skills to identify what data is useful and to avoid finding patterns that are meaningless. Most people are very good at finding patterns in data: it's why people can recognise faces. But some patterns don't always mean anything, they are just coincidences. And other patterns may not be something you can make a profit from. For instance, it may not be a coincidence that a large percentage of your customers are blue-eyed (because you have an effective sales operation in Estonia) but as you manufacture spoons this is not be a pattern you can use.

And you need the right reporting tools. Data visualisation tools allow the creation of innovative ways of displaying data and can also allow you to interact with that data in real time. This can be of great value to your board who, instead of requesting reports for the next board meeting, can generate the required information then and there.

Automation in finance

Automation involves machines (including computers) that are able to complete repetitive tasks at a fraction of the time it takes humans to do so and with greater accuracy and without complaint. There are a number of financial processes that may be appropriate for automation. These could include making payments, credit control and financial reporting.

Until now people have undertaken these processes. But if computers take them over, then people who simply record transactions do not have to be discarded. Instead, they could be offered a new role where they identify the value of information and help make decisions based on that information.

As we saw in the previous section, digital technology enables better analysis and better decision making. Automated processes can link up with data analysis and machine learning to make rapid and reliable decisions or to produce high-quality outputs such as reports. Automated analysis is increasingly better than human analysis. For instance, computers have been shown to be better at diagnosing skin cancer than doctors and can potentially select the best treatments. This doesn't make doctors redundant. The doctor is still needed to sense check the computer's diagnosis and administer the treatment and monitor results. The same is true with finance. Computers may well predict outcomes and identify risks. But finance professionals will be needed to decide how to react to those risks.

There are a number of benefits from automation for finance departments and these include:

- **Automated reporting**. Financial reports can be produced and distributed speedily and accurately. Businesses that use XBRL (eXtensible Business Reporting Language) and ESEF (The European Single Electronic Format for reporting financial results) can transform the way they can report their results, increasing accuracy, transparency and compliance.
- **Reduced costs**. Costs can be reduced significantly by automating key financial processes. In addition, there may be a reduced need to outsource specialist tasks with a saving not just in money but also in reliability and a reduction in the risks (including security) that come from using 3rd parties.
- **Better use of people**. Good talent is rare. And using it to undertake important but routine tasks that could be completed by a computer is wasteful for organisations, and frustrating for the talent. By freeing people up to do things that add greater value such as supporting financial decision making, organisations increase effectiveness and job satisfaction.
- **Increased speed**. Processes can be completed more speedily. Even relatively complex tasks such as reviewing contracts or auditing companies

can be conducted by computers. This reduces risk and opens more opportunities.

- **Improved quality**. Data processing quality can be improved with automation. There will be a considerable reduction in error rates compared with human data input.
- **Greater consistency**. Using a machine to undertake processes means that you can be sure that the process is conducted in the same way, whether it is in Paris, France or Paris, Texas. If financial consistency is put in place across different business units, it is far easier to compare performance.
- **Better records**. If a machine undertakes a task it is simple to record when and where the task was undertaken and who caused the task to happen. It is much easier to ensure control and traceability with automated processes.

As with data analysis, the automation of processes brings along with it certain requirements. In particular, there is a need to understand any human issues: the threat of redundancies needs to be handled and imagination will be needed when deciding how best to redeploy affected staff.

In addition, there is a need to have the strength of mind to accept that while the computer may be right a lot of the time, it isn't always right. For instance, software may have been programmed using the wrong data set. Or the software may malfunction in certain circumstances. And there may simply be a mechanical breakdown such as a power outage. There will be a need for constant monitoring and review by humans.

And finally, while automation is a great benefit, it won't be appropriate for every financial process, or for the whole of a process. An understanding of when judgement (i.e. the presence of a human) is needed in a process will always be needed.

Finance process integrity

But there is another advantage available to you, as a finance professional, from the use of digital technology. Software applications can be designed and developed to exhibit the characteristics of what might be called "integrity by design". Such features can include the enforcement of such things as:

- Account administration
- Account blocking and surveillance
- Authority limits
- Delegated authorities to roles or persons, typically as exceptions to normal rules
- Approval authorities, including exceptional, additional, authorities
- Processing priorities, e.g. against timelines or compared to other tasks or first calls on spending
- Audit sampling
- Budget escalations.

This technology capacity is presently often under-deployed even within existing capabilities. And that's important given the fact, reported by Action Fraud (1), that in 2017–2018 UK businesses lost £88m to employee dishonesty, with the average loss exceeding £62,000. Certainly, financial processes that use technology to provide integrity by design would have largely prevented the CEO fraud discussed in the box below.

The latent power of financial controls is also set to be multiplied many times as emerging technologies such as Robotic Process Automation (RPA), AI and other emerging technologies are used, many of which we examine further in Chapter 14.

Getting it right (and wrong)

For professionals who are used to trusting digital technology, using more to get greater financial insights might seem like a low-risk strategy. But care will be needed when implementing new digital processes.

You should ensure that you are confident that processes have been designed with sufficient rigour to deliver the required outputs. There is always a danger that the apparent ease of setting up new digital processing, or the apparent efficiencies they bring, leads to a loss of process integrity and consequently increased fraud. In addition, there will always be a need for common sense. Answers that fly in the face of experience may well be flawed – perhaps caused by an incomplete data set. Trusting what the computer says isn't always a good idea.

Unfortunately, all too often, people trust what computers tell them. In doing this they are of course delegating responsibility for their decisions to the people who wrote the software that the computer is running and the people who collected the data that the computer is crunching – people who may not be financial experts and who almost certainly are not intimately familiar with the problems that are being analysed.

This trust in machines, often but not always appropriate, is perhaps the single biggest disadvantage of digital technology in financial management. But there are other issues too.

COMPANY SUES EMPLOYEE FOR FALLING FOR A FRAUD

In 2015 finance worker Patricia Reilly fell for a "CEO fraud" which cost her employer Peebles Media Group over £100,000. Criminals had sent her emails that pretended to be from her boss. As a result, she transferred nearly £200,000 to a fraudulent account. Some of the money was recovered from the fraudsters but, considerably out of pocket, her employers decided to sue her for the difference.

They maintained that Ms Reilly had received an automated warning about fraud by the bank when she made the transfer and had therefore been negligent. They also claimed that she wasn't authorised to make payments on behalf of the company. Her boss was reported as saying "If I had known she had been on the banking system, I would have gone apoplectic".

In her defence, Ms Reilly claimed that she had been given no training on how to spot a fraud.

Two questions come to mind. If Ms Reilly wasn't authorised to be on the banking system, how did she get access? And if she was, how was she allowed to make such large payments without them being confirmed by another member of staff?

Appropriate technology use would have prevented access while a stronger "four-eyes" process might well have resulted in the fraudulent payment being stopped.

The increased use of digital technology extends the attack surface of an organisation, making it more susceptible to hacking and online crime and fraud.

The use of automated messages, or even email, may give the process designer or email sender a feeling of confidence that a message has been delivered. And it may well have been. But that doesn't mean it has been read or understood.

There may also be a temptation to overanalyse data when it is relatively easy to do so, at the expense of decision making: paralysis by analysis.

As we have said elsewhere, people are very good at detecting patterns, even when the patterns don't really exist or are merely coincidences. It is important that people have the skills to identify valuable patterns and ignore worthless insights.

Digital technology is undoubtedly a powerful tool for financial management. But organisational leaders must be confident that due care is taken to ensure that over-confidence does not creep in at the expense of common sense. They need to know that any decisions are taken based on the analysis of appropriate data, and that the use of automation technology as a short cut does not in reality generate extra and unnecessary work. And this is as much an issue for human resources strategy as it is for financial strategy, as the next chapter will show.

Reference

1. Action Fraud. (2019). *Young Staff Commit Most Fraud | Action Fraud.* [online] Available at: www.actionfraud.police.uk/news/young-staff-commit-most-fraud [Accessed 24 May 2019].

9

HUMAN RESOURCES IN A DIGITAL AGE

Summary

The HR department is being transformed by digital technology. This isn't because people are being replaced by computers. It's because a number of very powerful new ways of working are open to HR teams that make the most of digital technology. But as always with digital technology, many risks are present and these need to be addressed carefully.

Digital technology can help employers find the right employees although it can also, if badly used, help you find the wrong ones. It can make your organisation appear to be a more attractive place to work, strengthening the "employer brand". It can be used to manage employees and to strengthen culture. But perhaps most importantly of all, digital technology provides opportunities for individual employees to deliver greater value to their employers. Managing those employees so they have the skills and the motivation to do so is a critical challenge for organisational leaders. And the HR department itself needs digital skills.

Perhaps it is surprising, but the Human Resources (HR) department has been, and is being, transformed by digital technology quite as much as other more obviously digital functions such as finance and marketing.

Processes around recruitment have been strengthened by the ability of organisations to identify potential recruits, even when they are not actively looking to move (passive job seekers). Employers can use technology to increase skills levels, deliver fairer work assessments and provide more powerful incentives to stay. This is as true at the most senior level as it is for new graduates and other juniors.

However, there are inevitably a number of pitfalls that you need to guard against. These include:

- The ability of people to deliberately embroider (or conversely accidentally damage) their attractiveness as candidates or employees, as demonstrated in online information they post about themselves
- A tendency to value new digital skills, seen as cool or perhaps "something we probably need (if only I knew more about it)", over more traditional skills
- A temptation to see digital technology investments only in terms of cost saving and to justify digital projects by focussing on just delivering headcount cuts
- A failure of on-boarding, move and off-boarding processes to take account of digital technology and its associated issues such as security.

Given the importance of digital technology to HR, having enough digital competence within the department itself is essential. So too is ensuring that HR is being a beacon of digital good practice, not least as HR departments are so intrinsically tied up with facilitating due privacy of applicants and their personnel's information.

And across the function there is a need for HR to step up and accept that they have a vital role in managing how organisational culture is affected by digital technology, especially in areas such as bullying, maintaining skills and knowledge and remote working.

Senior HR managers must have a common vision of how they want a prospective employee to see their organisation, as well as how they are seeing it in practice, before they actually visit the offices or talk to someone on the telephone. The web, and particularly social media, is often the first taste a potential employee gets of an organisation.

Within more junior HR roles there must be sufficient awareness of, and competence in, how to use digital technology effectively, especially when posting information online, in the public domain, to identify and woo the very best talent.

Recruitment

Finding the wrong people

Finding the best talent is a major concern for most organisational boards. In a time of high employment levels, they may well be hearing from their HR teams that people with the right skills are impossible to find.

This may be true of course. There are well-established shortages of mid-wives in the UK, for instance, as there are cyber security professionals. But in many cases the shortages are created by the HR departments themselves. Why is this?

Let's start with qualifications: it is easy for recruiters to focus too much on the qualifications that ideal (or rather idealised) candidates have. But these qualifications may not be appropriate to a digital world where computers do much of what used to be done by humans.

Take engineering as an example. The skills engineers need include: creative thinking, attention to detail, teamwork, problem solving, communication skills and leadership. These are often largely absent from traditional academic courses, perhaps because they are hard to teach. But increasingly the traditional engineering skills involving mathematical calculations and computer modelling will become redundant as computers do the computational heavy lifting while humans do far harder tasks around identifying business requirements and then delivering projects successfully.

Of course, some roles need specific skills and experience. Midwives, for instance. But if recruiters insist on using filters that exclude people with non-traditional qualifications from roles that don't necessarily need traditional qualifications, then it is unsurprising that they find vacancies hard to fill. One example might be recruiting cyber security professionals. You might assume that all candidates need traditional network security management skills. But in fact some will need other skills such as psychological or textual analysis instead.

As technology changes the way we work, recruiters need to be constantly aware of changes to the way they should be selecting candidates. We should not be selecting people on their ability to compete with computers.

That's not to say that the mental agility and precision provided by studying mathematics isn't important. But that agility and precision could be just as well taught through learning a musical instrument or Mandarin Chinese. And the skills around communication and teamwork will be better taught in drama classes than in physics.

Another issue caused by HR departments is age discrimination, which is something that both younger and older workers experience. The excuse used is that the younger candidate doesn't have the knowledge while the older candidate has only out-of-date knowledge. What matters more is, do the skills that they presently have demonstrate their ability to adapt and to take on new skills? Digital technology provides us with instant access to knowledge about almost anything. It gives us the ability to learn constantly, throughout our working lives. It's true that people may need to be guided in how to find and apply the right knowledge. But a lack of specific knowledge should rarely be a reason for filtering out talent.

You may be tempted to filter out workers because you feel they don't have the ability to update their knowledge. And you may decide to filter out younger workers because you feel that they don't have enough experience. But accept that when you do this, you are showing prejudice to a particular set of skills and a particular set of people. Don't make knowledge an excuse for this prejudice.

Finding the right people

Digital technology can make it a lot easier to find the right people, especially those mythical creatures – the passive job seekers, the people you want to employ but who are not currently looking for a job.

You can often identify potential future candidates on social media, especially on business networks like LinkedIn, as well as Xing (Germany),

Maimai (China) and Viado (France). When there, it can be useful to look out for people who:

- Actively join in online discussions in popular and active industry groups, in a positive manner, starting discussion threads, commenting on other people's posts or even simply "liking" posts
- Regularly share useful industry news on Twitter
- Describe what they do, and what their employer does, clearly, correctly and positively on social media profiles
- Are not indiscreet on social media.

However, it is always sensible to be cautious about what people post about themselves. So here are a few more considerations when using social media to identify candidates:

- People often see social media as a place to boast; checking social media profiles against more formal CV documents is essential, as is taking up references and examining their posting activity.
- Posts made on channels such as Facebook and Twitter may not reflect the person's character or abilities. Many people who are aggressive or foolish on social media are perfectly pleasant and sensible in the flesh.
- Don't assume that someone with large numbers of social media followers, or a large number of LinkedIn connections, is in some way a "star": they may simply have bought those followers or applied no criteria to accepting connection requests.
- Openly using social media to identify potential candidates can lead to discrimination claims by unsuccessful candidates because there may be information on social media (e.g. age, gender, ethnicity) that won't appear on a standard CV.
- Some people are simply uninterested in social media, while others may have entirely valid personal or professional reasons for keeping a low social media profile – if you rely solely on social media you will miss these people.
- Attempting to link to someone on social media merely to find out more about them for the purposes of potentially recruiting them is likely to be a breach of privacy laws, especially if they have flagged themselves as not being open to speculative approaches.

ELIMINATING BIAS WITH ROBOTS

Recruiters can be biased for all sorts of reasons. For instance, gender, age and ethnicity can all mark you down, even if the recruiter is unaware that they are being biased. In an attempt to combat unconscious bias in the recruitment process, a Swedish recruitment agency TNG has experimented with using a robot called Tengai to conduct interviews.

There are of course problems with this approach. Some interviewees may well feel very uncomfortable to be interviewed by a machine and might not shine. Others will perhaps feel that using a robot shows the company is uncaring and would be a bad employer. And the robot itself is unlikely to be as good as a human at probing an interviewee's answers, even if they are (as is perfectly possible) better at picking up lies than a human might be.

Perhaps the biggest problem though will be ensuring that the robot doesn't display bias, as has been seen with several other robots powered by artificial intelligence (AI). Bias could come from the robot's programming or from the information it is given to build up its machine learning. (See the textbox on AI in Amazon on page 237.)

The "employer brand"

Is your organisation one I want to work for? Assuming I don't know many people who already work for you, I am going to answer that question by going online and by what I see in the general media. But I am probably going to start with social media.

Increasingly, rightly or wrongly, many people will assume that a company that is not active on social media is failing. A company without an active Twitter account uses out of date technology; it's overly hierarchical; it delivers products and services that no one wants to buy anymore. In short, it's headed for bankruptcy!

For that reason, it's important to think through and document how you will build your employer brand online. Even if your organisation is actively using social media, you should think about how you will promote your company as a good place to work. To do that:

- Use social media like Facebook and Instagram to build a positive image of what it is like to work for your organisation. Post (selected) images of company social events (don't name or "tag" people unless they have said OK) as well as other occasions that might show your organisation in a positive light (e.g. industry awards, charity races). Encourage employees to submit imagery such as videos (perhaps run an annual day-in-the-life-of competition where people create short films of their work environment).

- Identify any negative comments such as reviews on Glassdoor.com and engage with the complainant. Your aim should be to prevent further complaints from them. If they have a genuine grievance it will be best to do this offline (not least for reasons of privacy). Using the law is rarely a sensible tactic as you will come across as a bully (and therefore not a place where most people will want to work). By contrast, that individual being prepared to close their postings with a positive message about positive handling will be disproportionately beneficial to you.

Treating candidates as customers

Some elements of the job search process are tedious. How many times have you submitted a CV to a potential employer (or their recruitment agency) only to be told to fill out exactly the same information in an online form? How many times have you applied for a job only to hear nothing back (there is no real excuse for that in a digital age)?

If you treat candidates with the same care that you would treat customers, then you are likely to retain the best ones and retain the respect and custom of most of them. Digital technology can help here, for instance, recruitment platforms such as launchpadrecruits.com, which aims to give candidates an easy and engaging journey through the recruitment process. Enable candidates to communicate easily and informally, perhaps using online chat bots to answer questions or hosting group webinars where people can ask questions.

Remember, the average cost of hiring someone in the UK or the USA is about £3,000 – so treating them like customers (who probably cost a lot less to recruit) is sensible.

Managing social media reviews

Some social media applications, such as Glassdoor, explicitly target workers, asking for reviews of their experiences and their managers. Of course, it's fairly hard to manage this type of thing but there are some things that can be done. And as recruiting the best talent is something that most leaders are very concerned about, this may be something you will want to supervise closely.

- Create an account. If you create a Glassdoor account, then you can make sure basic company data is correct but more importantly, you can engage with people who are reviewing your company.
- Use the reviews. Bad reviews generally (but not always) happen because of bad management practices. Use bad reviews as a source of intelligence about the way your organisation is managing employees.
- Ask your employees, perhaps especially those who have just had a raise, to write kind reviews; this could work very well after a successful social event such as a Christmas party.
- Respond to reviews, and do so in a positive, rather than a defensive, way.

On-boarding

Once candidates have been recruited it's important for the board to know that contracts and instructions contained in employee handbooks cover digital technology appropriately, completely, clearly and unambiguously. It will be most important that the organisation has evidence of acceptance of these by a new recruit, or by someone who has moved roles and is being exposed to new technology, data or digital processes. However, a digital solution could be used to collect and store such evidence.

Questions to ask include:

- Does your policy about using IT and the internet at work cover the use of private devices like smartphones to access and store confidential company information?
- Do you claim and actively manage (e.g. keep copies of passwords) the ownership of any digital IP created at work? This is particularly important where employees create online assets such as websites or handle social media accounts as access to these may be hard to retain in the event of the employee leaving.
- Do you explain sufficiently well the new employee's responsibilities when it comes to cyber security, to the privacy of other people

(customers and prospects but also fellow employees) and to the resilience of the organisation during unexpected events?

- Do you emphasise the disciplinary actions you may take if the new employee fails to follow the rules, especially what actions are considered to be gross misconduct and therefore grounds for instant dismissal?
- Do you have a social media policy that explains when and how an employee may make reference to your company, your clients and your industry? For instance, you may think it is appropriate for an employee to claim they work for you on LinkedIn but you may not want them to do so on Facebook or Twitter where their posts about your company may appear next to posts about what they were up to on Saturday night.
- Do you advise new employees about whether, when, where and how they may (and therefore may not) connect with colleagues (especially line managers and junior colleagues) on social media?

It is important to have a slick on-boarding process with regards to technology (the new employee is all set up with the right – but only sufficient for their role – work accounts, equipment and access permissions the moment they arrive) that maintains the positive employer brand you have already built up. New employees should be given clean, or at least cleaned, IT devices (computers and work phones) that you are therefore sure don't contain confidential information relevant to the last user and their last role.

Digital technology isn't just a headache though. It can be used to help with the process of on-boarding, e.g. checking that policy documents have been read and signed, project teams joined automatically, and background checks conducted (prior, for instance, to access being given to confidential information).

Managing workers

Technology has a large part to play in the management of people within organisations. It should of course never be used as a substitute for personal interactions between a manager and their subordinates. But it can add some useful elements, such as performance data, as well as, inevitably, providing some problems that need careful handling.

Using data

HR professionals have an opportunity to collect and use data about employees in order to manage performance more efficiently. As well as streamlining analysis, it can be argued that the automated processing of data makes performance analysis less subject to personal bias. New opportunities come from sentiment analysis and natural language processing where text (in social media reviews, appraisal documents and internal emails) can be analysed for signs of dissatisfaction.

IBM, as might be expected, are experts at this type of analysis and claim to be able to spot 95% of workers who are about to quit, by using artificial intelligence (1). As well as enabling line managers to prepare for an imminent departure (or perhaps try to prevent it) this intelligence can be very useful for cyber security professionals looking to focus on the behaviour of those employees who may be more likely than their colleagues to steal confidential data before moving to a new employer.

Getting emotional at work

Skilled managers sometimes base decisions on three things: what people do, what people say and how people feel. But detecting how people feel isn't easy. We can look at their faces and body language and we may be able to detect a good deal. However, some workers are good at hiding their emotions. And some managers are poor at detecting emotions.

Machines have been used to detect emotion for some years. The reactions that people have to films, websites and adverts are sometimes analysed using data taken from facial expressions. Increasingly, machines are being used to detect emotion at work. This may be in a job interview, but potentially the technique could be used more widely, in appraisals, for instance, or even when people come through the door at the start of the day.

Creepy? Definitely! And you would have to question the ethics of such an approach especially if people have not been told that this sort of analysis is being undertaken. But is it effective? That probably depends on how it is implemented. Put people at ease and they are likely to act naturally. But tell them that they are being monitored in a formal situation and they are likely to respond, perhaps by exaggerating certain facial expressions, perhaps by hiding them. And of course, if they are in physical discomfort or feeling unhappy for some reason, that too is likely to result in misleading data.

Motivation

Flexibility over internet use

There is no right way to treat the use of digital technology and services during the working day for non-work purposes. Whatever rules are appropriate for your organisation, these should be clearly explained in policy documents that are shared with employees. Ideally these should be explained face to face to new joiners, and the penalties for not following them made clear.

Some companies forbid the use of social media and personal shopping sites, for instance, while others are perfectly happy to allow reasonable use or will confine use to certain hours such as breaktimes and after work hours.

The upside of forbidding non-work activity, or at least excessive amounts of it, during the working day are obvious. But so are the downsides:

- Many workers expect to be able to use social media throughout the day, just as they expect to be able to answer or take personal telephone calls. Preventing them from doing so may demotivate them and will certainly damage your employer brand.
- Giving people the ability to take regular short breaks from stressful work activities can increase productivity (2); denying them breaks can lead to errors or bad decisions.

The truth of the matter is that even if you forbid personal internet use during working hours, and perhaps prevent work computers from accessing anything other than a white list of authorised sites and apps, people are still likely to use their mobile phones to check out their WhatsApp messages and see what's happening on BuzzFeed. If you are not prepared to have that, your policies will need to have established that in advance.

Boards therefore need to define and communicate what their policies are and what they consider is excessive and/or unacceptable behaviour. And all of this needs to be consistent with whatever your policies are on using personal devices at work (BYOD).

USING TWITTER LATE AT NIGHT CAN GET YOU FIRED

In 2014, PayPal was forced to part company with a senior executive after he posted a series of bizarre and insulting late-night tweets including: "Duck you, Smedley, you useless middle manager" and "People who should be fire from PayPal Don Christmas a pool a kick".

Rakesh Agrawal blamed his tweets (but presumably not his inability to write coherently) on a new phone that he couldn't use properly, saying that he had intended the tweets to be private "Direct Messages" to a colleague (3)

Twitter is often seen as being an informal medium that somehow doesn't "count" and where content is ephemeral, disappearing into the ether after a few people have seen it. But what you post on Twitter certainly does have commercial weight, and does remain, as Mr Agrawal discovered. It also has considerable legal weight and people have gone to prison for making threats on Twitter (4).

Managing email sensibly

Email is the bane of most office workers' lives. You probably dread coming back from a holiday, in part at least because you know there will be hundreds (or more) emails to trawl through. The problem is that email has grown up without any real commonly understood etiquette.

If I copy you into an email, it is probably because I want to prove to you that I am doing something praiseworthy or because I feel it will help me to shift or blur the responsibility for a decision I am making. As a result, I am giving you one more unimportant email to read (or ignore). Surely, if I am looking for praise, getting it face to face would be more satisfying. And if I want to cover my back about a decision that I have made then copying you into an email won't provide me with much cover.

In addition, if you send me an email and expect me to reply to it immediately you are simply worsening the situation by strengthening the hold that email has on the working day.

We all receive far too many emails each day and spend far too much time on the (generally easy) task of reading and forgetting them. Developing a healthy email culture will involve a few, probably much appreciated, rules:

- Turn email notifications off and only review emails at set times, perhaps twice a day.
- Never copy people into emails unless there are justifiable (and legal) reasons for doing so. If they need to read your message put them in the To box. Automatically send emails where you are copied into your trash file (don't do this for blind copies or BCC emails though – someone probably has a reason for BCCing you).
- Set a limit for the number of emails people are allowed to send in one day, and make this part of their appraisal. Encourage people to talk instead, face to face or on the telephone.
- Encourage people to remember that words on their own, without context, tone of voice and body language, can easily be misinterpreted.

Some organisations have found it beneficial to institute "email-off" periods, replacing them with walking and talking instead.

Email culture may at first seem as if it is too trivial a task for your board to concern itself with. But when you realise that, by some estimates, office workers spend a quarter of their time dealing with them, then perhaps it starts to become a highly strategic issue that can affect the organisation's productivity and its success.

Incentives

Technology delivers new ways of working and delivering that can then conflict with traditional ways of working. One common problem is the way that online activities can conflict with or complicate offline activities.

As well as making sure that there isn't a conflict between different teams (for example, the digital vs paper sales conflict we mentioned in Chapter 3), digital processes should be seen as causing extra work, especially if that extra work seems to be unnecessary. Unfortunately, while new digital systems are being bedded in alongside existing analogue systems, some duplication of effort may be inevitable but people should be assured that this is only a temporary solution. It's where a new system seems designed to create extra work for no reason, or where it doesn't take work away when it could, that resentment will build. A customer database that uses two different email systems (perhaps one for people who have signed up to a newsletter and one for people who have not) may mean that operators have to do tasks in one email system that are not required in another: explaining why this is necessary (and perhaps apologising!) will be helpful.

Incentives don't always have to involve money. Simple recognition is often all that is needed and digital technology can make this very easy with online performance award nomination schemes capable of taking data in automatically and able to be shared easily around an organisation.

Saying "thank you"

Technology can be used to give instant appreciation to your colleagues. This can be as simple as posting a positive comment or an up-vote on a document or idea that a colleague has submitted on an intranet or innovation platform.

In addition, software can be used to provide recognition for simple-to-achieve key performance indicators such as keeping email storage down or contributing to the intranet. Of course, this doesn't replace a face-to-face thank you from the boss but the public nature of online thanks adds some considerable value.

Privacy

No one likes being spied on at work. And while it is legal, in the UK at least, to read the emails that employees send or to listen in on their phone calls, this is a right that should be exercised transparently and with due care. Certainly, this should only be after asserting that right at the start of employment and getting documented acceptance of the practice.

The UK's privacy watchdog, the Information Commissioner's Office (ICO), puts it like this (5): "Data protection means that if monitoring has any adverse effect on workers, this must be justified by its benefit to the employer or others". In other words, you need a reason for monitoring people, such as a suspicion of inappropriate behaviour. Simply saying, in one of those recorded messages, that this is for training purposes, unless there is a demonstrably-aligned training programme, is at best a cop-out.

Privacy is particularly sensitive when it comes to social media. It is one thing monitoring when people are on social media and quite another to monitor what they write. Of course, it will be hard to monitor what people say in private posts unless you are linked to them in some way (for instance a Facebook friend). Some people are tempted to suggest to subordinates that they "friend" them, and this request can be hard for the subordinate to say "No" to. This can be dangerous as it can lead to claims of spying or even bullying and discrimination.

The governing body needs to make clear to all personnel with supervisory or other higher responsibility what is and is not expected of them.

Health and safety issues

Digital technology raises some health and safety issues. While computers have got a lot lighter over the years (fewer back injuries!) there are still some issues to consider.

The need to provide appropriate ergonomically designed furniture and to encourage screen users to take breaks and have regular visual checks are probably well known. However, connectivity has changed the way we work, and not always for the better.

Being always on is pernicious. Expecting colleagues, especially junior colleagues, to reply to work phone calls and emails outside office hours is tantamount to bullying. But worse, it is bad for morale (and therefore employee retention) and for productivity. There is considerable research (6) linking long working hours to lowered productivity. This is because long working hours are bad for health: they cause stress and anxiety, reduce the ability of the employee to take exercise or eat well, and damage social and family life. The temptation that many senior managers have to keep working when they get home, and keep their colleagues working too, is therefore likely to be self-defeating as well as uncivilised.

As well as being itself pernicious, it also tends to lower the quality of the work actually done. A dashed-off reply to an email, because it was received on a smartphone at 21:00, on a Friday, is rarely if ever well-considered or complete. Even if it is, it probably won't be perceived to be.

As we said, forcing people to work long hours is bullying. But bullying doesn't only come from the top down. It can be seen between peers and is often facilitated by social media. Colleagues who fall out may, if connected on social media, continue their quarrels on Facebook or Twitter. This can lead to claims for discrimination with the employer held vicariously liable.

Off-boarding

Ultimately every employee leaves the organisation they are working for. And when they do it is important to have an off-boarding process that takes account of digital technology. In particular, passwords to online services such as social media and software subscriptions should be obtained and then changed so that they are no longer usable by the ex-employee.

One thing to consider here is file-sharing applications. If your employee has been using a service like Dropbox or Google Drive (presumably with the organisation's approval) to share files with colleagues, then access to those files needs to be rescinded. Of course, if they have been sharing files through their own private Dropbox account, this might be difficult. In which case, a declaration by the soon-to-be ex-employee, that they now do not hold or have any access to any of the organisation's information, should be obtained. This would be vital in any subsequent legal recovery that is attempted. But sensible organisations that allow Dropbox use will surely have set up a corporate account that they control.

Many organisations fail to do this, however. Some 68% of workers store work-related information in a personally managed file-sharing solution. And the vast majority (84% in the case of Dropbox, 68% with Microsoft SharePoint) continue to have access after they leave (7).

It will typically also be appropriate to remind the departing employee that they are no longer authorised to say they work for you (we are assuming you have this stipulation in one of your policies or in their contract) and so profiles on LinkedIn, Twitter and anywhere else you allow this information to be changed should be changed immediately, perhaps as part of the off-boarding session itself. In addition, remote access to any corporate information, e.g. via home-based VPNs, should be cancelled.

Enhancing the human

Automation

Many people fear that digital technology is inevitably going to cause mass unemployment as computers and machines take over jobs that people use to do, and do them faster, better, more safely and cheaper. And there is no doubt that some jobs will be automated away.

That's nothing new. Automation has been happening for several hundred years and during that time, in the UK at least, employment rates have increased rather than decreased. Of course, technology will inevitably cause a shift in the nature of employment. In the United States, farming jobs made up 40% of the workforce in 1900, but by 2000 that share had dropped to 2%. Similarly, 25% of jobs in 1950 were in the manufacturing sector, but by 2010 manufacturing was less than 10% of the workforce, according to a McKinsey Global Institute report on automation and employment. In both cases, new jobs, sometimes in new industries, eventually offset the losses.

That's not to deny that technology will cause short term challenges as old skills become redundant, and some people find it hard to reskill. But automation, as it gets rid of dull, dirty and dangerous jobs, providing people with jobs that create more value (generating higher wages and less need to work long hours) is likely to benefit society greatly. We talk more about the opportunities that automation will bring in Chapter 14. Nonetheless, a challenge for society, and employers, remains – ensuring that the current workforce and the workforce of the future have the skills they need to flourish in a digital economy.

Digital literacy and skills

If you accept that digital technology underpins organisations today, then you should also accept that today's workers need to be digitally literate. Unfortunately, this acknowledgement is not always the case.

It might seem counter-intuitive, given that many school leavers seem to have their smartphones surgically attached, but many people entering the workplace are digitally illiterate. They may have the ability to do certain things such as play games or music, but they are unable to use digital technology to undertake work-related tasks such as filling in forms or finding reliable information.

Schools are clearly failing in their duty to educate children about digital technology. An employer who takes this responsibility on itself would therefore be strengthening their workforce and making it more fit than those of its competition for today's requirements.

The International Society for Technology in Education (8) suggests that learners need to demonstrate digital literacy through a number of skills that employers could help to develop. These include:

- **Empowered learning**, e.g. the ability to use technology to learn. This should be easy for any organisation to help with as there are numerous learning resources available free or paid online.
- **Knowledge curation**, e.g. the ability to locate trustworthy resources online. This is an essential skill for anyone in the workplace tasked with researching a subject.
- **Computational thinking**, e.g. the ability to collect and analyse data. Related to knowledge curation, this is another essential skill for many workers, as well as being an essential life skill.
- **Creative communication**, e.g. the ability to communicate complex ideas using digital assets. Most workers need to communicate either with their immediate team members or with third parties and using technology to do this, whether orally, using text or using images, can be very efficient.
- **Innovative design**, e.g. the ability to use technology to solve problems. This is perhaps harder for employers to deliver as many employees won't be tasked with solving this type of problem; however, they can encourage employees to research solutions to common societal problems as a way of illustrating this (9).
- **Global collaboration**, e.g. the ability to use collaborative digital technologies to work in teams. This can be experienced through team exercises where people work together on the design of a new product or a new way of doing things, using file-sharing and other collaborative technology. And this is most definitely not the same as taking part in multi-player rounds of 'Call of Duty'.
- **Digital citizenship**, e.g. the ability to manage your online identity. Advising new, and existing, employees about how to manage their identity online, and particularly how to keep it safe from identity theft, would be beneficial to individuals as well as teaching them valuable lessons about cyber security that can be applied at work.

These are all skills that any organisation will benefit from. But they need to be backed up by appropriate knowledge. And, with technology changing so rapidly, the knowledge that workers have must change rapidly too.

Grasping the essential truth that business skills acquired yesterday will be obsolete tomorrow is frightening. And it requires a sea change in the way that many organisations treat training and skills development. It's no longer sufficient to employ someone with a particular set of skills and expect these to be relevant several decades (or even several years) later. Life-long learning is required.

Many professionals will feel that they are safe here. Their professional status requires a degree of Continuous Professional Development. And this is obvious for lawyers or accountants who need to keep up to date with changes to the statute book and for medical professionals who need to be aware of the latest techniques for addressing illness.

But it isn't just professionals who need to keep up to date. A mechanic whose job involves tightening nuts with a wrench needs to know how to use (and perhaps how to request and justify) a smart wrench that enables exactly the right amount of torque to be applied. A journalist needs to know how to record and edit a podcast and how to promote it on social media. And a lorry driver of 20 will almost certainly need to accept that their job will have ceased to exist by the time they are 40 and that they may have a future career, using their existing skills, as a drone pilot (hand-eye coordination), a security guard in a casino (constant alertness for the unusual), a counsellor (ability to manage stress and loneliness), or doing something you simply cannot foresee at the moment.

For all of these people, and everyone else, flexibility, a willingness not to be defined by your job and a hunger to learn new skills constantly will be essential. And this attitude of mind is just as important for organisational leaders, who must promote the culture that supports this openness to technology, as it is for everyone else within your organisation.

Knowledge management

Knowledge (not data as some might say) is the lifeblood of organisations.

Knowledge can be defined as an understanding, gained by personal experience or tuition (someone else's experience), of the things that are happening in and to an organisation. It is the word "experience" here that is key: knowledge is information *plus* experience. And experience is personal. Your experience of reading this book is very different from mine. And both of us had very different experiences of arriving at work today.

Experience is very useful because if you share it you can help people understand what might happen under a particular set of circumstances. Because of this, it is much more useful than mere information which only tells you "what, when and who". In contrast, knowledge can tell you "why and how" as well as "what, when and who" and might even tell you what to do next.

Information is seductive. It is easy to store and share, easy to manage. But it isn't much use. Knowledge, because it includes personal experience, is far less easy to manage but it is much more useful. Therefore, organisations need to think about how they can collect and curate knowledge. And you won't be surprised that digital technology can help.

Because knowledge involves personal experience it is internal or tacit. The first thing you need to do is to change it from tacit to explicit by getting the person who has that knowledge to share it. People can make knowledge explicit in a number of ways: through talking, drawing, writing or demonstrating and ideally a combination of these.

Encouraging people to turn knowledge from tacit to explicit is a skill in its own right. Not only do people find it difficult or wearisome to describe their experiences, they may well be motivated to hide them for a number of reasons including fear and greed (information – or in this case knowledge – is power).

When knowledge is being made explicit it needs to be captured. For instance, if you ask a colleague to describe how they won a piece of business they might simply write you a case study but more likely they would talk to you about it. And in that case, you would need to capture what they say. Digital technology can help in two ways here: it can turn spoken words (sound) into written words; and it can help with the analysis of written words in a number of ways:

- Identifying the words (including names) that were used
- Identifying the sentiment or emotions that were expressed
- Providing some analysis of content such as the dates and places the events described occurred
- Providing some analysis of the speaker from body language, tone of voice and even typing style.

There is nothing very new about the analysis of natural language. But powerful computing is making our ability to analyse words ever more powerful and accurate, avoiding confusions caused by slang and irony and enabling the meaning of paragraphs, rather than just words, to be extracted.

Once the knowledge has been made explicit, it can be curated – sorted, prioritised and described so that other people can find it and use it. Again, technology can help here, for instance, with sorting and describing knowledge, although prioritising its importance effectively is likely to require humans for some time yet.

And finally, the explicit knowledge can be shared over digital communications networks anywhere you choose in the world.

Digital culture

The culture of an organisation is largely set by the governing body and its top management team. By identifying the vision and the values of the organisation, and then displaying them in their personal performance, they will extensively influence what people believe about their work and how they too should behave.

A cavalier attitude to cyber security, for instance, will be reflected in a weak security culture and an increased risk of cyber breaches. A dismissive attitude to technological developments will stifle innovation. And a reluctance to invest in new technology will result in frustrated employees who lose trust in the abilities and potential of the organisation they work for.

Cultural issues though go beyond whether your organisation is seen as a shiny new place to work. Digital technology can impact organisation culture in many ways and governing bodies need to be aware of this and, where necessary, act to move culture in the right direction, not least by setting an example.

Some issues you may want to consider are:

- The use of personal devices such as smartphones and tablets in meetings. Are so many people really taking notes or are they answering emails and text messages?
- Restrictions on the use of the internet at work for personal activities such as shopping and social media. You may well feel it is appropriate to ban these activities, but they are likely to happen anyway on people's smartphones. Far better to try to restrict them to sensible amounts of use.
- The stressful "always-on" culture of work.

It's a truism to say that for most organisations, people are the most important asset. But organisations also need to accept that new ways of managing people as they use technology are needed. The relationship between employees and technology is a complex one. There are issues of knowledge, usability and motivation to address. Perhaps for organisational leaders, the people challenge is the greatest challenge that digital technology brings.

References

1. CNBC. (2019). *IBM Artificial Intelligence Can Predict with 95% Accuracy Which Workers Are About to Quit Their Jobs.* [online] Available at: www.cnbc.com/2019/04/03/ibm-ai-can-predict-with-95-percent-accuracy-which-employees-will-quit.html [Accessed 24 May 2019].
2. This article gives some reasons why this may be true: Psychology Today. (2019). *How Do Work Breaks Help Your Brain? 5 Surprising Answers.* [online] Available at: www.psychologytoday.com/gb/blog/changepower/201704/how-do-work-breaks-help-your-brain-5-surprising-answers [Accessed 24 May 2019].
3. Mail Online. (2019). *New PayPal Exec Goes on Angry, Poorly Spelled Twitter Rant.* [online] Available at: www.dailymail.co.uk/news/article-2619622/New-PayPal-exec-goes-angry-poorly-spelled-Twitter-rant-calling-colleagues-useless-piece-s-t.html [Accessed 24 May 2019].
4. Marsden, S. (2019). *Twitter Trolls Jailed for Abusing Feminist Campaigner.* [online] Telegraph.co.uk. Available at: www.telegraph.co.uk/technology/twitter/10595669/Twitter-trolls-jailed-for-abusing-feminist-campaigner.html [Accessed 24 May 2019].
5. Information Commissioner's Office. (2019). *Quick Guide to the Employment Practices Code.* [online] Available at: ico.org.uk/media/for-organisations/documents/1128/quick_guide_to_the_employment_practices_code.pdf [Accessed 24 May 2019]
6. Eric Robert's CrunchMode team at Stanford University have explained how a 60 hour working week delivers only two thirds of the productivity that a 40 hour working week does. Crunch Mode. (2019). *The Relationship Between Hours Worked and Productivity.* [online] Available at: cs.stanford.edu/people/eroberts/cs181/projects/crunchmode/econ-hours-productivity.html [Accessed 24 May 2019].

7. Osterman Research. (2014). *Do Ex-employees Still Have Access to Your Corporate Data?* [online] Available at: www.intermedia.co.uk/assets/pdf/do_ex-employees_still_have_access_to_your_corporate_data.pdf [Accessed 24 May 2019].

8. Panda Security. (2019). *45% of Ex-employees Continue to Have Access to Confidential Corporate Data - Panda Security Mediacenter.* [online] Available at: www.pandasecurity.com/mediacenter/security/employees-confidential-corporate-data/ [Accessed 24 May 2019].

9. Some simple examples of how technology can be used to solve problems are given here: EDUCABA. (2019). *22 Amazing Ways to Solve Problems with Technology (Simple).* [online] EDUCBA. Available at: www.educba.com/how-to-solve-problems-with-technology/ [Accessed 24 May 2019].

10

ASSURING DIGITAL COMPLIANCE

Summary

Compliance is a key issue that most organisations will be concerned about. But compliance isn't just about keeping on the right side of the law (and we don't discuss laws in this book as they vary from nation to nation and require specialist advice). While organisations need to do certain things to remain lawful, many compliance obligations are self-imposed, for instance, compliance with international standards. And this is because compliance can deliver many benefits – reduced costs, better quality products, lower risks, greater market authority.

Being compliant isn't simply a question of following a process that has been laid down by others diligently (although that helps). It requires organisations to want to comply, and as part of that to comply with the spirit as well as the letter of what they are complying with. It requires them to use common sense and take an approach based on what is reasonable. It requires them to understand what they need to comply with, for instance, existing and emerging industry standards. But compliance needs to be built into organisations with regular audits to assure compliance, rather than treated as a one-off tick-box exercise, if the full benefits are to be enjoyed.

Owning and running a digital technology estate was never an easy thing to do, but it is set to get harder – at times seemingly hour by hour. There are calls for ever more regulation, especially in respect of digital services, often driven by concerns about privacy abuses, the publication of fake news or the (perception) of bad practices by social media companies. This represents a real change in expectations – whether that is coming from the general public, your customers, your operating partners, regulators, the government or international bodies.

The challenge that comes with not taking digital compliance seriously (or not appearing to do so) is the perceptions you generate with your customers, employees and others of poor, unethical or even negligent behaviour. They will then typically respond with reduced loyalty, causing sustained damage to your reputation and brand. Some say that the days of simple trust are over. Certainly, trust is hard won and easily lost, as no end of recent scandals show.

In considering compliance and assurance, you need to address both those obligations you may have had placed on you by your partners and obligations that you have imposed upon yourself as part of delivering your own mission and goals, doing so in an integrated way.

DISCLAIMER: COMPLIANCE WITH LEGAL OBLIGATIONS

In writing this book, we thought long and hard about covering the compliance obligations that arise from the law. But we ultimately concluded that this would be unhelpful to you. We aren't by any means expert in this area. It is complex, extensive and fast changing, it varies considerably between jurisdictions and our commentary might do considerable harm to you. We appreciate too that it is not always about just what the law says, it is also about how it is being interpreted, what precedents have been set and what current cases are being progressed in the courts.

We strongly recommend that your digital governance plans include regular input from Legal Counsel, whether that be an in-house team or from an external legal practice. They will also be able to advise you

> about when client privilege may apply, or can be made to apply, to a case. This convention essentially means that you have legal rights to refuse to disclose, and to prevent any other person from disclosing, confidential communications between you and your lawyers. This can be very helpful, even if it might only give you breathing space to consider your responses, in claims of liability, etc.

It is also important to recognise that any non-compliance can move rapidly into the justice system and may well result in much stiffer penalties than was previously the case.

Why care about compliance?

Digital compliance is too often treated today as a synonym for the EU's General Data Protection Regulation (GDPR), observing the USA's NIST cyber security standards or achieving an ISO27001 certification. But compliance issues are far wider and more nuanced than digital privacy and cyber security. (We consider those topics in appropriate detail in Chapters 11 and 12.)

Sadly, despite the increase in legislation that applies to digital operations, sometimes triggered by a simple lack of awareness, compliance is too often written off as onerous, as an overhead upon your organisation's operations, all cost and no benefit. However, in stark contrast to this, there is both increasing evidence and acceptance that your effective implementation of required, and/or best, practice is genuinely value generating; even if it were only to make your organisation stand out from the crowd.

The wider benefits beyond customer confidence and loyalty, that apply as much to digital services as any other part of your organisation, include:

- **Increased authority**. As one of the good guys, you may well get to punch above your weight when dealing with others, such as regulators.
- **Remaining within a community**. Meeting expectations of behaviours and continuing to be able to use a service by being recognised as a trustable third party, derived from a confidence in your operations.
- **Management of risk**. Controlling your exposure to risk by implementing effective controls and optimising your appetite for it.

- **Reduced operating costs**. Generating increased margin or increased service delivery through the implementation of established best practice.
- **Good relationships with regulators**. If regulators are confident that you are employing best practice then they are less likely to investigate you and more likely to give you a chance to address any issues they find if they do investigate you. And if you do get a fine then it is likely to be less than the maximum possible, as the regulator will know that you have done many of the right things.

Sources of compliance obligations

Compliance is certainly not simply about you meeting a regulator's expectations. Compliance is first and foremost about meeting the expectations you have set yourself as part of meeting your organisation's strategic objectives and goals.

You may well have chosen to impose upon yourself an obligation to meet a range of formal standards, either in their own right or as a means to the above ends, which we discuss further later in this chapter. However, many of your compliance obligations will have been placed upon you by others. Again, the range and sources of these are numerous and will include requirements that you have:

- As membership of an industry body who have collaborated to specify a common baseline or level playing field, e.g. electronic document submissions in the legal arena.
- As a contractual condition of utilising a service. For example, in 2017, SWIFT (the global provider of secure financial messaging services) imposed on its members a Customer Security Programme to establish a series of security baselines in order to protect the integrity of those services.
- As a contractual condition of a contract you have entered into, e.g. to provide secure, confidential and resilient processing services.
- As part of national law and regulation, e.g. the Computer Misuse Act 1990.
- As part of international law and regulation, for example, the EU's GDPR.

Perhaps you should be forgiven for any (false) perception that compliance is just something for the finance industry. Indeed, the extent of compliance obligations now applicable to that industry is already very extensive and still growing. Examples of regulation in this sector include:

- Local regulations such as the UK's Financial Conduct Authority's distance selling regulations and in the USA the SEC's social media regulations
- Regional regulations such as the EU's Payment Services Directive
- International regulations such as SWIFT's Customer Security Programme.

SOCIAL MEDIA AND COMPLIANCE

Gene Morphis, CFO of Francesca Holdings, a boutique fashion company, was an enthusiastic social media user. Unfortunately, this habit led to his dismissal. A series of posts, while appearing on the surface fairly uncontentious, led to this. After an audit committee meeting, he wrote on Facebook "Damn you Paul Sarbanes! Damn you Michael Oxley!" (a reference to the US congressmen who had promoted the 2002 financial regulatory law. And around the same time, he tweeted "Board meeting. Good numbers=Happy Board". The post was from his Twitter account which was fairly anonymous as it was called "theoldCFO". Unfortunately, the account linked to his Facebook account where his role as Francesca's CFO was highlighted. This was considered to be the inappropriate disclosure of financial information and was enough to get him fired.

However, it's hard to think of any sectors that have no compliance obligations. The challenge, whatever your industry, is to recognise that your use of digital technology, as it gets ever more central and fundamental to your operations, becomes yet another domain in which there are compliance obligations which may include defining, sustaining, demonstrating and documenting certain digital behaviours.

For example, building construction requirements to meet building controls, such as bright strips on the leading edge of stair treads particularly for those with eyesight limitations, are well understood. But the digital analogy might be to highlight important text on a website and to make data entry fields intuitive and easy to use.

In another example, in the first quarter of 2019 in the UK, there were calls for the regulation of organisations involved in transgender work – whether counselling, treatment or after-care support. The early discussions were mostly about person-to-person interaction, but such organisations are inevitably using digital technology to facilitate and increasingly shape their work, in which case those organisations will need to extrapolate the regulation requirements, especially privacy, onto the digital systems they employ.

Becoming compliant

Scanning and tracking expectations

All of the above demonstrates that your organisation really needs to resource and maintain a process of:

1. Scanning for emerging compliance obligations
2. Tracking their content and their impact upon your organisation
3. Mapping those into the digital technology controls that are consequently required.

It is commonly the case that regulators judge that an organisation is doing a good job of compliance just by instigating this process.

Furthermore, digital technology that is implemented without any consideration of compliance needs, or without functionality that can support it, will undoubtedly damage your organisation. This will be especially true when the implementation is by non-IT professionals.

You will do well to ensure that implementation teams for new digital technology combine business process, IT and compliance experts within inter-disciplinary teams.

Engagement not denial

If your scanning, tracking and mapping are good, then it has to be even better for your reputation if your organisation is involved in the creation, shaping and proving of new compliance obligations.

Governments, trade bodies and professional institutes are all constantly running consultations and bodies like the British Standards Institution (BSI) are always deriving standards, and especially about the topic areas in this book (for example, controls over AI, distributed ledgers, robotic process automation, see Chapter 14).

The cost of engaging in this way may occasionally be significant but the payback – in terms of defining, piloting and early adoption – will be greater still. Furthermore, it is then likely that the effort needed by your organisation to fit itself for the new compliance will be reduced while the quality and effectiveness of your compliance will be better overall.

Obligation or option: what's reasonable?

Some of your compliance activities will be prescriptive in their execution, but many won't, and even the law uses such phrases as "adopting acceptable, proven digital controls frameworks" to limit that proscription (or to avoid admitting to the limits of their knowledge).

Those that are not prescriptive are therefore, to a greater or lesser extent, optional. And if not prescriptive, then their blind application is to be avoided where possible. Having decided you want to comply with them, you need to think about to what extent as well as how.

Overall you should want to meet the spirit of the compliance requirements. Avoid your staff thinking in terms of boxes to be ticked.

A growing trend within compliance obligations is to set objectives to be met rather than being highly specific about the practices to be followed. For example, the EU's General Data Protection Regulation includes some specific requirements for particularly sensitive issues but otherwise is founded upon a series of rights that must be upheld along with clear discretion about how they are then delivered.

This is analogous to the accessibility requirements in the UK's Disability Discrimination and Equality Acts, where taking *reasonable* steps is required. It would not necessarily be reasonable to require a small bed and breakfast or an upstairs café to install a lift for disabled customers. In the same way, websites need only to be reasonably accessible to the disabled if they offer services to them.

What may or may not be reasonable will vary in different fields and scenarios, but it will help if you can show that you have:

- **Worked through the requirements**, typically against a good range of scenarios of accidental and/or deliberate breach, e.g. poor application design.
- **Logically and carefully selected your responses** with a clear rationale, e.g. through a digital risk register.
- **Documented your logic and decisions** in a policy document or similar, e.g. a digital development policy.
- **Trained your staff in the subject** and kept a record of that training having been done, e.g. via an e-learning facility.
- **Kept the subject under review**, documenting instances of breach or near misses and learning lessons from them, e.g. through structured software testing against each of the application design objectives.
- **Maintained effective reporting**. Whether or not there is a regulator to whom compliance reporting is due, your organisation's governing body's ultimate accountability means that it should be receiving regular, periodic, reports on compliance status, e.g. as part of a submission to permit application deployment.

An increasingly common practice of regulators, when they are obliged to think in terms of fines, is to consider the non-compliant organisation's behaviour in terms of this lifecycle, with percentages of the theoretically possible total fine allowed being allocated to different steps – poor commitment, lack of preparation, lack of training or poor reporting, etc. Your defence will be your assessment of your organisation's risks, your records and those who are ready and willing to speak in your support.

The role of standards

Compliance requirements are being increasingly defined as objectives, enshrined in law but without the details of how to do it defined. The result is a demand for formalised standards to fill that gap.

The best example of this is occupational health and safety. In the UK, this came to the fore with the 1974 Health and Safety at Work Act. Most organisations now expect to comply with the relevant international standard, ISO45001, to meet the expectations of their stakeholders. There are many comparable standards that apply to the digital domain.

We owe you a word of warning about formal standards. Their gestation follows a carefully regulated process that establishes a common baseline, one that reflects an extensive consensus. They therefore take time to produce and will always lag behind early learnings. The plus side is that standards generally have considerable validity. This is surely better than simply reaching out to peers to see what they can advise. What trust can you place in their advice if there is no empirical evidence of its efficacy?

The standards universe

There are standards and then there are standards. Many informal standards are generated across a membership, a sector or a profession; if you adopt these you are gaining from a consensus of knowledge. But there are also formal standards. These are derived at national, regional or international levels, for example:

- **UK standards** from the British Standards Institution (BSI). The UK has a proud record of generating standards that have gone on to be adopted as European and international standards. Many are standards against which formal certifications can be obtained.
- **European standards** from the Comité Européen de Normalisation (CEN), the European Committee for Electrotechnical Standardization (CENELEC) and the European Telecommunications Standards Institute (ETSI). The EU Commission has recommended that these bodies merge but, even now, there are 60,000 subject matter experts at work (including those from the UK of course).
- **International standards** from the International Standards Organisation (ISO) and International Electro-technical Commission (IEC). Here the subject matter experts will number in the thousands and standards will take at least 3 years to be created but are then the ultimate in consensus of good practice.

Management system standards

Many of ISO and IEC's standards take the form of operational process or lifecycle models for the sustained conduct of an activity. Many of these standards, in turn, have frameworks for accreditation or certification associated with them. These frameworks then deliver formal recognition of your efforts in the given space. The most common example of this is the certification logos seen on the back of vans on the motorway. Your implementation of these management systems will come at a cost, well outweighed by the benefits derived, and are perhaps almost the ultimate in standards. These include the following existing or planned standards:

BS13500 – Organisational Governance
ISO9001 – Quality Management
ISO14001 – Environmental Management
ISO22000 – Food Safety
ISO31000 – Enterprise Risk Management
ISO37000 – Organisational Governance
ISO37001 – Anti-Bribery Management
ISO45001 – Occupational Health & Safety
ISO55001 – Asset Management

But perhaps not quite the ultimate. As well as some fundamentals (including digital governance) not yet standardised to that degree, there is also a limited, but now rapidly growing, level of standardisation of how to integrate multiples of these systems together for efficiency, consistency and effectiveness.

Emerging compliance obligations

As discussed previously in this chapter, the creation of new compliance obligations is very much à la mode. Chapter 11 considers the EU's Networks and Information Systems Directive (NIS) and, in Chapter 12, the EU's General Data Protection Regulation (GDPR) is described.

Another key area to watch is the intention of the UK, USA and other governments to regulate social media. After growing public concerns over the failure of the media giants' processes of self-regulation, the UK Government published proposals in April 2019 to put in place external social media regulation requirements, expected to be implemented via legislation. These proposals appear to be consistent with intentions within the EU and the USA (1).

The proposals are designed to "establish a new statutory duty of care to make companies take more responsibility for the safety of their users and tackle harm caused by content or activity on their services". Under them, an independent regulator will be given powers to fine organisations and individuals or block sites delivering online harms. The funding would come from a levy on social media companies. The fines that could be levied for non-compliance are likely to be substantial, similar perhaps to those of the GDPR where 4% of global turnover is available as a sanction for the regulator.

The harms were defined as terrorism, child sex abuse, so-called revenge pornography, hate crimes, harassment, the sale of illegal goods, cyber bullying, trolling, fake news and disinformation. Surprisingly, there were no proposals about revisiting age limits nor about removing self-rating by the social media companies. One challenge to be resolved is that terms like disinformation have yet to be formally defined.

SOCIAL MEDIA: AN UNCONTROLLABLE CHANNEL

Many, in fact most, organisations feel that it is highly desirable to use social media to market themselves. Sometimes there is a desire to have a "viral" marketing campaign. Viral marketing involves commercial messages like adverts that strike a chord with members of the public who then share that message, perhaps because they think it is funny or clever, with other members of the public, freely spreading your adverts across your target audience and probably a lot wider.

While there can be considerable upsides to social media, such as this viral effect, there are also many risks. The problem is that social media is impossible to control and very difficult to predict.

Kylie Jenner was just 21 in 2019 when she was crowned by *Forbes Magazine* as the world's youngest dollar billionaire. The reality TV star had a cosmetics company then worth over £750 million, over 130 million followers on her Instagram account and 25 million Twitter followers. And she was previously an avid user of Snapchat, another social media messaging platform. But after a change to the design that she disliked, she tweeted, "sooo does anyone else not open Snapchat anymore? Or is it just me ... ugh this is so sad". The tweet was liked over 250,000 times and caused an 8% drop in the share price of Snapchat's parent company.

In social media there really is such a thing as bad advertising, it would appear.

Social media unpredictability can work in other ways. In early 2019, the web editor for a chain of local radio stations in Texas came across a story that he found interesting and published it. The story had a worrying headline: "Suspected Human Trafficker, Child Predator May Be in Our Area" (2). Surprisingly, this local story spread widely across social media in a short period, garnering over 800,000 shares on Facebook.

Why did it happen that a story intended for a local audience received such prominence on Facebook? It's probable that because the story seemed interesting ("Child predator") but had no precise geographical location ("in our area"), Facebook showed the story to a far wider set of people than it would have had the story limited its geographical appeal with a headline including the words "in Texas").

Whether or not there is a case for regulating social media, it's certainly important for organisations to accept that the medium is highly uncontrollable and that using it presents a number of risks that they need to be prepared for.

Although coming with considerable difficulties for effective enforcement, there seems to be extensive support for the move.

But there are also real challenges. It is unclear what organisations will (and then won't, of course) be in the scope of the legislation. Some of the harms are hard to define and thus ill-defined at present. For example, what constitutes disinformation? Would it apply to anti-vaccination campaigners who consider themselves to be using their right to free speech? If it did, would they be obliged to remove their content and how will that be ensured? Will these global businesses run the risk of being fined 3 times over (or more over time) for the same offence? And who will sort out their online content if or when they go bust?

There is a more direct set of potential downsides for your organisation. Your digital services could be considered to put your organisation in scope. Your personnel's misuse of the digital realm could mean that your organisation is vicariously liable, as it looks likely Morrisons will be judged to be for the cyber breach of their systems (see Chapter 12). Your organisation might even end up as collateral damage if the social media giant that hosts your digital services is found in breach.

Whilst the abuses of social media continue, and the legislation beds in, you and your business need to give the development of this regulatory framework close attention.

Rationalisation of compliance

It is highly probable that, across the various standards and requirements with which you are required to comply, that there will be areas of duplication. In some cases, one of the duplications may be mandatory, and you won't want to lose sight of it.

But often, the different requirements are driving at the same objective for broadly similar reasons, i.e. there is no specific driver for the particular wordings used. In such cases, you are strongly advised to invest the effort to produce a rationalised set of requirements.

These so-called "Integrated Controls Frameworks" can typically be supported on regulatory technology or RegTech platforms (see later in this chapter) that support an efficient "comply once: report many" approach. This reduces your overall compliance effort whilst, at the same time, boosting your compliance attainment and confidence.

Challenges of digital technology compliance

The IT profession has been with us for some 80 years, but still feels like a cottage industry at times. It is perhaps fairer to say that despite the creation of numerous and extensive standards, those standards have not been comprehensively implemented and not therefore fully validated.

Digital technology standards

This position is, thankfully, changing at pace – with considerable focus upon it, and standards production by – the standards bodies described above. The following standards underpin many of the concepts outlined in this book and especially the framework described in Chapter 2:

BS10010 – Information Classification, Marking and Handling
ISO19770 – Information Technology Asset Management
ISO20000 – Information Technology Service Management
ISO22301 – Business Continuity Management
ISO27001 – Information Security Management
ISO27014 – Governance of Information Security
ISO28001 – Supply Chain Security Management
ISO38500 – Governance of Information Technology for the Organisation
BS31111 – Cyber Risk and Resilience

Even using these standards, there are particular challenges with the actual deployment of digital compliance. These relate to whose job it is and to what extent "full compliance" (an extremely dangerous term) can ever be claimed.

Digital technology is used right across the organisation. So digital compliance involves the whole organisation. It certainly is *not just* the responsibility of the IT Department (itself a concept that is visibly dis-solving) and of their supply chain partners, even though they may be well placed to provide some of the day-to-day leadership on it that is clearly necessary.

Governing bodies need to ensure that everyone sees digital governance as a part of their remit. This may involve directing behaviours, specify-ing controls, operating those controls, reporting back on that compliance, learning the lessons of non-compliance or approving new behaviours. We have discussed in Chapters 8 and 9 how this needs to be reflected into your organisation's finance and HR activities. But the dynamic nature of digital activities and the pace of development of new technologies also suggests that you should *never* claim that your organisation is fully compliant with regards to the various requirements of digital technology. Instead, your focus and statements should be about being as compliant as it is reasonable and practical to be today, with the determination to work to maintain and develop that position in the years ahead.

This type of statement is clearly more in the real world than any claim of full compliance and moves the onus onto the other party to prove different. And this is especially sensible when you think that standards are themselves the work of flawed human beings and are not necessarily perfect.

TARGET USA: APPARENTLY COMPLIANT, BUT HACKED ANYWAY

Target, a large USA retailer, experienced a painful cyber security breach in 2014 (3). This breach exposed personal and financial information of more than 110 million customers.

The criminals hacked into the Target system using access rights stolen from one of Target's suppliers, a heating and air conditioning contractor called Fazio Mechanical Services. The hackers were able to gain access to the stores' Point of Sale (POS) equipment which held the credit card details of customers.

This breach raised questions about Target's compliance with the Payment Card Industry Data Security Standard (PCI DSS). If Target was compliant to the PCI DSS, as they claimed, then how was the data stolen? The hackers should not have been able to move from the systems that Fazio were able to log into right through to the tills in the stores.

Clearly, no standard can provide absolute security. But the breach raised questions. Was Target fully compliant? Or was the standard in some way flawed? Either, or both, of these may be true. Certainly, the Payment Card Industry updated their data security standard shortly after the breach.

Doing what you said you'd do

You've said what you will comply with and you've set expectations and given direction to top management. They have worked with you to define, establish and operate appropriate controls. You believe you have carried that through effectively. But what have you got to show anyone that challenges you, when after all you carry the final accountability?

We discussed earlier that everyone has a role to play in delivering compliance, but that should really be extended to include the assurance of compliance.

Aspects of digital governance such as this can usefully deploy the Three Lines of Defence model (4) originally developed by the audit community and initially within the financial services sector (Figure 10.1).

In this model, the first line of defence is the management of the operations themselves conducting what might be considered self-assurance, i.e. checking

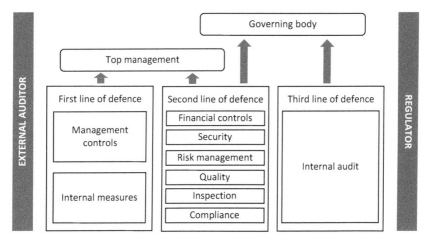

Figure 10.1 The three lines of defence model.

upon themselves. This of course fits well with requirements for regular reporting of operational compliance flowing ultimately to the governing body.

The second line of defence is then assessments and audits conducted upon the operation by internal specialists such as risk management or compliance departments. They too report of course to the governing body.

The third line of defence is then activities performed by independent third parties such as internal or external auditors, also reporting in. In some sectors, the risk profile and process criticality is such that there are more lines of defence (e.g. adding peer reviews into the second line of defence or treating external auditors as still second line defenders and having external consultants as the third line of defence). But whatever model is employed it clearly delivers strong assurance.

Assessments, audits, KPIs and reporting

Governance is often described as a boardroom-to-storeroom process. If it is, then it is clear that achieving assurance requires both top-down and bottom-up activity.

You should assess your compliance and benchmark this against your peers. You may well do that by scoring yourself against a set of KPIs that you've agreed upon, for instance, using a traffic light model or a set of metrics that will be tracked.

You should make these arrangements subject the results of the Three Lines of Defence model if you are operating this. Together, these give strong confidence in, and justification of, your chosen compliance stance and attainment status.

Assuring the assurers

If you decide to include external parties in your assurance framework, select them with care, ensuring that they have the competences in the field required. It is not a given that even the so-called Big 4 auditors have all the requisite skills. Certainly, their names on a report can have a powerful weight in your favour with regulators, etc. but they are still human and have been shown to not be immune from compliance failures.

Furthermore, for yet more checks and balances, it may well be beneficial to explicitly not ask your normal external audit partner to do this work for you. The report from an additional third party will help to assure the effectiveness of one of your assurance functions too.

The benefits of assured compliance

The assurance of your compliance is not, first and foremost, about protection in the case of regulator or criminal or civil court action. It is about continuous operations improvement.

But continuous improvement is not automatically about "more" and "gold-plating of" requirements. If your compliance level was established with an eye to commercial pressures or levels of risk then it may be that your compliance assurance findings concern the over-development of compliance, the increase in cost that you were worried about.

You should ensure that the findings from compliance and assurance activities are as often reviewed for what can be dropped as for what more might be needed. The world outside is changing, so you could sensibly include some horizon scanning in your periodic compliance reviews.

Third-party reports can have a very strong impact upon any enforcement activities of a regulator, especially if a sanction is being considered. With that in mind, you should ensure that the subject matter expert you are employing as an assurance reviewer has a contractual obligation to speak to the regulator in support of their report to you.

The rise of RegTech

The rise of compliance activity, assurance needs, Three Lines of Defence and benchmarking have undoubtedly combined to create an explosion of reporting data to manage. This has triggered the advent of specific regulatory technology (RegTech (5), as it is becoming known). This is a dynamically growing market space and one that will require its own effective and commensurate digital governance. But it is one that is capable of facilitating higher levels of compliance and digital governance at a considerably lower cost.

Regulatory technology pulls together regulations, standards and best practice guidelines to deliver analysis about compliance in formats ultimately useable by governing bodies. At its simplest, it involves the digitisation of manual reporting and compliance processes. At the more complex end, it can be about groups working to define objectives, develop plans and execute actions whilst collecting data and other evidence of attainment.

Applications include:

- Business strategy
- IT strategy
- Cyber security and information security
- Compliance management
- Customer due diligence and identification, e.g. the finance sector's "KnowYourCustomer" obligations
- Anti-money-laundering, anti-fraud measures and suspicious activity reports
- Auditing
- Risk management
- Supply chain management
- Transaction reporting.

In our opinion, it won't be that long before organisations will be criticised if they are not using RegTech to at least a minimum level to manage some or all of these areas. One in particular is likely to benefit: information security, a subject we discuss in the next chapter.

References

1. www.reuters.com/article/us-britain-tech-regulation-idUSKCN1RJ0QP [Accessed 6 August 2019]

2. Savage, A. (2019). *[CAPTURED] Human Trafficking Suspect Arrested after Brief Chase.* [online] KTEM NewsRadio 14. Available at: ktemnews.com/te xas-rangers-believe-human-trafficking-suspect-child-predator-may-be-in -our-area/ [Accessed 24 May 2019].

3. McCoy, K. (2019). *Target to Pay $18.5M for 2013 Data Breach that Affected 41 Million Consumers* [online] Available at: eu.usatoday.com/story/mon ey/2017/05/23/target-pay-185m-2013-data-breach-affected-consumers /102063932/ [Accessed 24 May 2019].

4. The Institute of Internal Auditors "Three Lines of Defence" principles are described at: Institute of Internal Auditors. (2013). *The Three Lines of Defense in Effective Risk Management and Control* [online] Available at: www.theiia.org/3-Lines-Defense [Accessed 24 May 2019].

5. For more on RegTech in financial services, see: FCA. (2019). *RegTech.* [online] Available at: www.fca.org.uk/firms/regtech [Accessed 24 May 2019].

11

INFORMATION AND CYBER SECURITY

Summary

Hacking and other cyber threats are increasingly recognised as of strategic importance to organisations. It's not just personal data about customers or employees that can be affected by a cyber breach. IP can be stolen or corrupted. Internet-connected machines can be damaged. Personal reputations can be destroyed. Large sums of money can be lost. The best talent can be deterred from joining.

Managing cyber security requires a highly structured approach that reaches across the organisation, not just the IT department. It needs to embrace people at every level as well as stakeholders such as suppliers, contractors and even customers. It needs to start with people – ensuring they have the right knowledge as well as a culture of security. It needs to address processes, ensuring the right rules and guidelines are in place, supported by management processes and monitoring. And finally (but only after people and processes have been addressed), it needs to ensure that appropriate technology is in place to support security. It is only organisational leaders who can make sure this holistic approach is effectively implemented.

As organisations automate and create more and more information, the security of that information becomes increasingly important and at the same time increasingly vulnerable. And increasing amounts of legislation and regulation require it to be secure, to protect customers, consumers, the general public and even the fabric of nations. But what has perhaps brought the subject to the forefront of media reporting and public interest is the seemingly continuous occurrence of security breaches with enormous numbers of records being leaked, stolen or lost.

The need to secure digital technology and to keep digital information safe certainly makes this a strategic issue and a key focus of a board's digital governance.

Information and cyber security are undoubtedly instances where an organisation's digital governance efforts need to translate to the engagement of the whole organisation and of the organisations in your supply chain.

In the UK, since about 2015, a strong focus of government has been kept on the professionalisation of security and increasing the flow of individuals into the profession to meet demand. This reflects a widespread concern about a real skills shortage. Recently, a senior executive showed us a report that claimed, of a group of students, 84% of them had stated that they wouldn't want a cyber security career.

But is that necessarily a gloomy outcome? One can infer from it that 16% would consider a cyber security career. And as there is a broad consensus that about 15% of an organisation's spending on digital technology and services should today be being spent on security it would seem there will be the resources necessary to meet this enormous challenge, provided that a consistent and structured approach is taken.

What is all this fuss about?

Cyber security breaches can cause real harm. And not just to organisations: to people too, whose lives can be disrupted if their identities are stolen or if services they rely on are dislocated.

DATA BREACHES AFFECT PEOPLE AS WELL AS ORGANISATIONS

In May 2017, the WannaCry ransomware caused considerable damage around the world to computer systems running Microsoft Windows. Of course, not every computer system was affected. The vast majority of computers had updated Microsoft software that was immune from the attack. But a substantial number of computer systems had unpatched (out-of-date) software that let the WannaCry software in. Two hundred thousand computers locked their users out and demanded payment of a ransom to let them get access to their data.

In the UK, one of the more significant victims of this attack were parts of the National Health Service (NHS). Nearly 10% of family doctors were hit and a third of UK hospitals. As a result, 19,000 appointments had to be cancelled at considerable human cost.

Whilst some might argue that the actual level of threat is not as bad as the media would have us believe, there has still been an exponential growth in cyber security attacks and failures. Why is this? More and more data is being created and aggregated. And there is more and more connectivity, both inter- and intra-organisational. These two factors have led to an expanded threat profile with a larger attack surface (there are more places where people can try to penetrate your organisation) that is harder to defend (because the connections between data are more complicated).

An often-used expression is: "An attacker needs only to be successful once, but the defender needs to be successful all the time". There's certainly some truth in that. Cyber criminals have an increasing level of motivation to learn, and to build capability as more data becomes available and defending it gets harder. They are building capability more rapidly than legal organisations and their security professionals. And they are typically better and faster at sharing and collaborating too.

At one time, the effort to secure an organisation went into creating a strong perimeter, turning the organisation into a digital armed camp. But the days of secure corporate IT perimeters have passed. Increasingly, people bring own digital devices (phones, tablets, watches) to work. People work from home or while they are travelling. And businesses use cloud computing (in effect using someone else's computer to store your data and then accessing it over the internet), and outsource more and more IT services. All these changes have made corporate IT perimeters almost invisible. Information security now has to work in different ways.

What's in a name?

For the last part of the 20th century, most people referred to the securing of company data, digital systems and services as *information security*. This had the advantage of being focussed upon organisational activity and outcomes rather than technology. And it allowed non-technologists to feel included in the discussion.

Since then we have seen the predominance of the term *cyber security*. We don't like this term but it reflects a number of factors such as:

- The complexities of modern digital technologies
- Their increasing pervasiveness
- The consequent complexity of cyber attacks
- The consequent complexity required of the technologies being used to prevent and defend us from those attacks.

Unfortunately, the term cyber security sends the (wrong) message: the subject is predominantly about technology and is exclusive to experts. This is worrying in that it largely excludes the consideration of accidents, especially those caused by the system's users.

Looking at the problem holistically

But all is not lost if, whatever the subject is called, you look at the problem in the round.

First, consider the many different *threats* to your digital services. These are a complex mix of internal and external, technical (e.g. computer failures), human (e.g. operator or configuration error), physical (e.g. fire or flood) and process (e.g. unauthorised use of an application). They are a mix of the accidental (user error) and the deliberate (hacking or data theft by employees). Their variety and volume are made worse by the speed with which they can become endemic very shortly after being seen for the first time "in the wild".

Next, consider that there are many different *threat actors* (the people who cause you problems) out there:

- Spies who commit corporate espionage on behalf of your competitors
- Criminals interested in the financial value of your data theft or intent on making money from ransom
- Hackers who are working for political purposes (hacktivists) or simply for personal pleasure
- Hostile nation states attacking you in the interests of their political aspirations
- Disaffected employees or other stakeholders who want to take revenge or perhaps undertake fraud.

These threats will, in turn, exploit different instances in a wide range of different *vulnerabilities*. These can be a simple lack of resilience (e.g. you only have one server dedicated to this task), a lack of capacity (e.g. a server is not big enough for the load it may have to process under certain circumstance) or deficiencies in hardware or software (e.g. the deficiencies perpetually being found in the operating system's software code) (Figure 11.1).

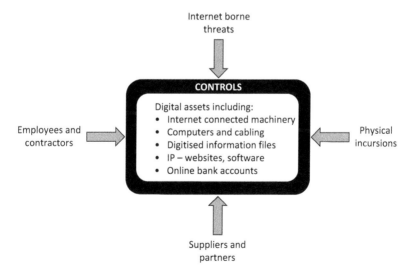

Figure 11.1 A simplified model of information security.

And finally consider that, though these threats and vulnerabilities combine to have their own direct effects upon different parts of your digital systems, they ultimately deliver *impacts* upon your organisation's business processes by causing reductions in your information's confidentiality or its availability and even its integrity.

If you recognise how these risks could impact your business, and if you recognise the realistic worst-case scenarios, calibrating them against financial and non-financial (e.g. reputation, employee morale, stakeholder relations) damage, you will have a chance of managing those risks, prioritising them and ensuring appropriate resources are allocated to their mitigation.

Developing a strategy

When the extent and nature of the information security task are understood, it's the role of organisational leadership, in association with top management, to develop an information security strategy. This will involve:

- Identifying the key information that the organisation holds, agreeing on the appetite for risk to these assets from a commercial and ethical perspective (remembering that there is no such thing as zero risk), assessing the nature of the risks to them and identifying approaches for managing those risks.
- Reviewing the areas where the organisation has to comply with laws and regulations, and the areas where it wishes to comply with standards and best practices; again, risk appetite must be established, risks identified and approaches agreed.

Cyber risk management

Major cyber risks

While there are many complications in cyber and information security that have come to prominence, there are two that perhaps merit consideration by governing bodies: phishing (because it is so common); and ransomware (because it is potentially devastating). You should ensure you consider both of these threats in your risk assessments.

Phishing

Phishing is the fraudulent attempt to obtain sensitive information such as usernames, passwords and credit card details by disguising yourself as a trustworthy entity in an electronic communication. It has a close cousin – vishing (voice phishing) – which is simply doing the same over a telephone. This social engineering technique tries to get users to enter personal information at a fake website, the look and feel of which are identical to the legitimate site, or to click on a link or file in an email such that malware (like ransomware, see below) can upload itself and operate.

Originally, many of these attacks were quite crude with spelling mistakes and fairly obvious lies offering instant riches; but mailed out to thousands of people they relied on the principle of "there's one born every day". These days most phishing attacks are very highly crafted and aimed at individuals (these are called spear-phishing attacks). And some of these will be directed at senior personnel in organisations like yourself (when they are called whaling attacks), people who have greater access, bigger budgets and more authority. Clearly, you owe it to your organisation and yourself to be highly sensitive to this threat and to act with care.

A note of caution though. It is often felt that training people to avoid phishing attacks is the answer to the problem. While training is useful it won't stop all attacks getting through. And in addition, anti-phishing training is often used as a proxy for the wider cyber security training, awareness and cultural change that is needed to guard against internal cyber security incidents.

Advanced persistent threats

Phishing attacks are commonly used in Advanced Persistent Threats (APTs). A typically impenetrable term, drawn from the sensitive side of governments, APTs are examples of technically complex threats (advanced threats) that can involve hackers loitering within your IT systems for extended periods (persistent threats) with the aim of waiting for valuable information to become available and then stealing it. Attackers can also route their attacks through your organisation to get to your bankers (for example) and your major customers, causing you collateral damage such as embarrassment at the breach along the way and you can ultimately suffer more directly, e.g. financial losses, whilst your customer's systems are down and then perhaps loss of the supply contract.

Once upon a time if you were discovered to be the route for such an attack you might have been able to protest your innocence and be excused. Not any longer.

Ransomware

Ransomware is malicious software that when inserted into a computer on your network will encrypt many of the files on that computer, or indeed across the whole of your network. Your IT system can be rendered useless until a ransom is paid, generally in a cryptocurrency like bitcoin. (The criminals behind these attacks generally provide very helpful instructions enabling you to make this payment.)

Ransomware is a very common problem for organisations. Many will opt to pay the ransom although that is ethically a dubious response and practically not always an effective one as there is no guarantee the criminals will actually restore your files.

The solution is relatively simple: backing up your data so that if the working copy is destroyed by criminals (or indeed other events) it can be restored. Data backups are a fundamental element of cyber security and it is as well to be aware of a few guidelines:

- Back up regularly: Data created between the time of the most recent backup and the present is at risk: how long this period is will depend on your risk appetite and the time criticality of the operations, but many organisations back up continuously to avoid any risk here.
- Back up in depth: Create several different types of backup. A continuous backup may simply import malware into your backup so it may be sensible to make a separate backup of your data on regular occasions (such as weekly and monthly) so that you can go back to this if your continuous backup is corrupted.
- Store backups in different places: If all your backups are stored on-site and your office premises suffer a power outage (or worse) then your data backup will be inaccessible. Storing it in the cloud is common and sensible but many organisations will also store a backup (perhaps of the most critical data) in a separate physical location.

- Test your backups: You cannot know that your backups will restore your data unless you test them. Waiting for a ransomware attack to see if your systems work isn't advisable.

Performing careful risk assessment

The activity described above is that of undertaking a risk assessment. As discussed elsewhere, especially in Chapter 2, you should take steps to ensure that this is done collaboratively with a wide range of stakeholder representatives. Ideally, you will do it in conjunction with other activities such as privacy impact assessments (see Chapter 12), information classification and the assessment of your resilience requirements (see Chapter 13) (Figure 11.2).

Once information security risk assessments have been undertaken, the governing body of your organisation should be presented with the analysis, to allow you to agree on the results, to satisfy yourselves that the organisation's mission, goals and objectives have been correctly reflected into those results, and to set, or adjust, your appetite to accept any of these risks.

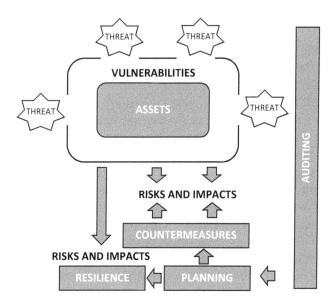

Figure 11.2 A simplified risk and response model.

Accessing the knowledge of subject matter experts is likely to be time and money very well spent. Ideally, at least some of your organisation's Non-Executive Directors will have experience in this area.

Risks in the supply chain

Information security does not end at the boundary of your organisation. You should be confident that your stakeholders, especially your supply chain partners, are secure. This consideration should start when you begin negotiations with them, it should be included in any contracts (which should mention the expected level of security and any procedures you require them to follow) and you should monitor their adherence to security service level agreements (SLAs) throughout the life of the contract.

How can you be sure your existing suppliers are secure? They might be relying on their brand reputation. But you ought to be looking for such things as kitemarks, the application of known standards, and whether they are already notifying you of breaches they have experienced and what they are doing about them.

A SPECIAL CASE: THE NIS DIRECTIVE

Within the EU, certain organisations have a particular responsibility to protect their information infrastructure: those organisations that are responsible for running a nation's critical infrastructure.

The EU's Networks and Information Systems (NIS) Directive (1) came into force at the same time as the General Data Protection Regulation (GDPR) which rather overshadowed it. But in terms of cyber security, the NIS directive is at least as important if your organisation (of at least 50 staff and/or a turnover of €10 million) is an operator of any services vital to the security of the EU and the well-being of its citizens – typically referred to as critical infrastructure. This could include telecommunication companies, banks, logistics companies, pharmaceutical companies and even some retailers. Breaches of this directive can attract fines of up to £17 million, rather than the 4% of global annual turnover under GDPR, but it will attract additional action by other regulators and by national governments too.

The directive adopts a layered approach to ensure that there is a national strategy for NIS and that NIS operators will have implemented *appropriate and proportionate security measures*. The directive includes four high-level objectives:

- Appropriate organisational structures
- Proportionate measures
- Capabilities to ensure defences remain effective and detect cyber security event
- Capabilities to minimise the impacts of a cyber security incident on the delivery of the essential services

These objectives could be met by using the structured risk assessment and layered approaches described in this chapter and they will also benefit from compliance with recognised controls frameworks, such as ISO 27001.

The NIS is probably a sleeping giant: its significance will only be recognised when an organisation is heavily fined (and publicly embarrassed) for a failure to comply.

Layered protection: people, process and technology

The risk assessment we talked about earlier is vitally important; but you've only done it as a means to an end – that of managing and mitigating the risks. You then need to put protection in place to manage those risks. And that protection should have three separate layers: people, processes and technology.

1. People: Set out your expectations of behaviours; for example, you might have a policy that employees must behave responsibly with data and that all passwords will be of a given level of complexity. You may well have other protections here such as employee education or stronger rules about mobile working.
2. Processes: Next address the ways that people are able to work, the processes within your organisation. For instance, you might prevent people from accessing certain parts of your network, such as your customer database, when they are working from home.

3. Technology: The final layer would be technology: for instance, software that prevents people from accessing dangerous websites, etc., as well as the proper deployment and operation of IT (Figure 11.3).

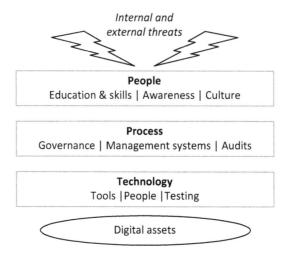

Figure 11.3 A layered approach to information security.

These controls should be consistently applied to your digital services. Typically, you should start by thinking about how people can be made to behave in more secure ways; then think about how you can make your processes stronger and less liable to be compromised; and finally (and it should be finally, rather than the first thing considered) think about whether there is a need for technology to bolster the security of your people and processes.

Controls for layered protection

A useful approach to ensuring you have layered protection in place is to identify the nature of a proposed control. The following list shows a series of levels, each of which has progressively less mitigation effect. The control types are:

*   **Avoid**, i.e. give up the activity creating the risk
*   **Transfer**, e.g. pass the activity to someone else, such as an insurance company

- **Prevent**, i.e. block the associated threat by a change to your people, processes or the technology you use
- **Reduce the likelihood** of the threat
- **Reduce the vulnerability** of the digital asset to the threat
- **Reduce the impact** of the threat on the digital asset
- **Detect** the damaging event caused by the vulnerability being exploited
- **Recover** from the damaging event and learn from it.

Categorising controls like this can significantly support related disciplines, particularly of privacy and resilience, but also of legal and regulatory compliance. The intention is to have a range of each of these in place, not to select just one.

The growth of considerable regulation (see Chapter 10) around the protection of digital assets may be of direct benefit to your organisation. They all either mandate the adoption of existing sets of controls or have specified their own requirements. These controls are likely to be useful across the whole of your organisation, although there will be a clerical task of combining them and removing any duplicated or overlapping controls.

Efficiency vs security

Keeping information secure is itself not a core strategic goal for most organisations. Effective operation is generally far more important: there is, for instance, little point in running a commercial organisation at a loss purely to keep information safe.

There is almost always going to be a trade-off between security and efficiency. As an example, insisting that people encrypt every document they create might be good for security, but it would cause enormous disruption to simple processes such as emailing colleagues or sharing draft reports. And overzealous rules can undermine the security they are designed to support: insist that everyone uses 18 random character passwords and people will simply write them down and stick them to their desks!

Any security processes need to take into account the value of the information they are protecting and the cost of implementing the process. And the cost of implementing those processes can be substantially reduced if attention is paid to the people who have to comply with them. Processes that "get in the way of the day job" and make it difficult for people to complete work tasks will at best cause frustration and at worst simply get ignored.

Managing assets that you don't control

One third of the world's population, around 2.5 billion people, have a smartphone that can connect to the internet. In Western Europe, this figure is around two thirds. And given that these devices are used for anything including telling the time, listening to music, using social media and playing games (and occasionally making a phone call), it's hardly surprising that most people bring these devices to work with them. This is often known as Bring Your Own Device (BYOD).

Bringing a smartphone to work isn't a problem unless you worry about people wasting office hours updating their Facebook status or doing their shopping. The real problem arises when it is connected to an organisation's IT network with the aim of accessing corporate data and completing work tasks. If this happens then data security problems can arise:

- Data stored on the phone can be taken out of a secure workplace environment; if the device is lost then the data may be leaked.
- If malicious software has accidentally found its way onto the phone, this might move onto your organisation's IT network when the phone is connected.

However, some organisations encourage their employees to use their own devices (not just smartphones but also tablets and laptops) for work purposes. This enables flexible working, with employees able to access and respond to emails while travelling between meetings or even work on documents when they are working from home. If this saves money and makes people happy, why not? There are, however, a number of considerations that need to be addressed at a senior level.

- **Paper trails:** Can activities undertaken on personal devices be audited later.
- **Safety standards:** Can their use compromise the security of corporate IT systems by acting as a doorway for malware.

- **Access control:** Will unauthorised people be able to access confidential corporate information on them.
- **Taking data out of the office:** Can they be used, deliberately or accidentally, to take confidential data out of a secure office environment. Are they secure when outside the secure office environment?

If personal devices are forbidden on an organisation's premises or IT network, perhaps as a way of strengthening cyber security, then it may be appropriate to allow access to certain websites via machines owned by the organisation, perhaps at restricted times or for restricted periods.

Personal software

As well as bringing their own devices to work (BYOD), some people will "Bring Your Own Software" (BYOS). There are some problems that arise from using software that has not been sanctioned by the organisation.

- **Security:** If insecure software on online services are used then corporate information may be put at risk.
- **Inefficiency:** If my colleague uses one software programme to create a document and sends me the file to alter, I may find it difficult to do so when I am used to using another programme.
- **Piracy:** If my colleague is using a pirated version of software, my organisation may be vicariously liable for IP theft.

Protection through people

Setting the rules

All employees need to have an understanding of how they are expected to behave when it comes to security. This understanding can be delivered through policies that establish rules of behaviour and explain why that behaviour is important. Policies should cover areas including:

- Computer and internet use
- Social media use

- Physical security
- Working with third parties and visitors.

Ideally, policies need to be explained face to face to new employees and the sanctions for non-compliance explained. A failure to do this may reduce the ability of your organisation to enforce these policies.

Training and awareness

Most people don't have an instinctive understanding of how to behave in a cyber secure fashion. The fact that, at the time of writing, the most popular password was still 123456 is evidence of that. Helping people understand how to keep information, including their own personal information, safe is the first step.

Policies are useful but security training is also important. Unfortunately, it is often treated as a tick box exercise with intrusive annual online tests that people have to pass (or, too often, get their assistants to pass for them) if they are not to be disciplined by being forced to attend more training. There are more engaging and creating ways to transfer knowledge.

But even when people do have the right knowledge, they won't necessarily act in a safe way. The pressures of everyday working life mean that security sometimes gets forgotten or simply ignored. It is important to keep people constantly aware of the reasons that behaving securely is important. Annual training won't do that.

Cultural change

Perhaps more important than knowledge and awareness is motivation. After all, most of us know what the speed limit is when we are driving; but how many of us can say we have never deliberately ignored it? Sometimes we are simply not motivated to do so.

It is the same with information security. We know what we should do but we don't do it because we are too busy or because we see other people behaving unsafely, or because we don't think it makes any difference. Ultimately, we don't think it is important.

And that's a cultural issue. Over time, developing a culture where behaving safely is "how we behave round here" is the only real way that you can build secure behaviour into your organisation. And that culture will largely come from the top. Your behaviour will have a massive impact on the security of your organisation.

Protection through process

Policies

Organisations need to develop and implement policies designed to protect security. These policies may include policies aimed at the organisation as a whole such as the incident recovery policy, which defines how the organisation should respond to disruptive events and disasters including specifying roles and responsibilities of the board and the wider employee base.

In addition, there are policies aimed at employees (including it should be noted members of the board):

- Acceptable use: Designed to provide guidelines on what is acceptable when employees use corporate IT systems.
- Identity and access control: A fundamental protection is limiting the access that people have to systems and information on those systems.
- Information security: A policy designed to provide guidelines on how employees should keep corporate information safe.
- Remote access: Guidelines on accessing corporate systems remotely, e.g. from home or a hotel.
- Social media: Guidelines on how employees should talk about your organisation, your industry, your colleagues and other organisational stakeholders when they are using social media.

As stressed earlier in this chapter, policies merely define the theory of what an organisation wishes to see. They have little use on their own and need to be explained to the people who will follow them. Monitoring the degree to which they are adhered to is also necessary, as are processes for taking action based on non-compliance, such as disciplinary actions or attempts to change culture.

Standards and best practices

One of the significant challenges facing cyber and information security is the rationalisation and formalisation of standards. At present, the (repeated) regeneration of standards is all too common.

Whilst it is by no means the only formal security standard around, the most commonly adopted, the most proven and the most dominant is the ISO 27000 series. This is a family of standards, about 70 in number, which address almost all aspects of delivering cyber and information security successfully.

The core standards in the family are underpinned by a certification scheme, through which you can get both third-party assurance and public recognition for your positive efforts. This public recognition gives security the same importance rating as quality, health and safety and environmental management.

Some may feel that the ISO 27000 standards are too big a first step. If so, there is a lighter entry scheme, *Cyber Essentials* (2) (sponsored by UK Government) that has a far lower entry cost. It comes with a much lower degree of assurance, but does still provide a public certification both of individuals and of organisations.

Operational processes and management systems

Senior security managers will no doubt have specified an appropriate security architecture that defines processes such as how and when software is updated, how access to corporate information is managed, how and when data is backed up, and how malware and unwelcome visitors (hackers) are kept off the network. These systems may well be in accordance with existing standards such as ISO 27001, Australia's *Essential Eight* (3), or the UK's *Ten Steps* (4).

But on top of these, there is an opportunity to ensure that particularly important or vulnerable processes are governed by extra controls. These may include:

- Making online payments over a certain amount or to new bank accounts, where a second set of eyes may be mandated
- Sharing sensitive data with third parties such as suppliers or contractors
- Setting up and managing public-facing digital assets such as websites and social media accounts.

In addition, it is important to control the access to corporate information of people who are new recruits, changing roles, leaving the organisation or even, or perhaps especially, people who are undergoing disciplinary procedures.

Embedding security with a Secure-by-Design philosophy

If you have undertaken board-level risk assessments and ensured the implementation of layered protection, this suggests that you will have made security an embedded feature of your organisation, one where:

• Security risk appears, however much summarised or aggregated, on your strategic risks register.
• Key (information and other) asset owners have been trained in security.
• Security is accepted as a cross-organisational issue, e.g. the HR Director recognises their key role in the fight.

This embedding can however be taken a step or two further. When developing new digital services, having a "secure by design" philosophy reduces risks from the outset. During staged development, through concept, design and development to testing, piloting and implementation, security consideration and security deliverables should be mandatory before progressing onwards.

But this concept ought to be applied to all other aspects of your operations too, whether that is outsourcing, off-shoring, entering new markets or offering new services. You should also schedule periodic security testing – penetration testing or at least scanning for known vulnerabilities – at a frequency that is broadly consistent with the identified level of risk.

Audits

Regular auditing is an important part of cyber security and governing bodies should take an active interest in their results.

Auditing can take the form of internal checks, such as provided by the UK's *Cyber Essentials*; or commercial service, such as provided by Huntsman Security's Scorecard (5), which ask a series of questions about an organisation's cyber security protections and profiles. Others, such as that provided by SecurityScorecard.com, provide an external view of an organisation, identifying vulnerabilities based in part on threat intelligence (for instance, whether hackers are talking about an organisation on the Dark Web).

In both cases, a score is provided which indicates where improvements can be made, and which can be used to show progress (or otherwise) over time.

Cyber insurance

The last few years have seen the growing emergence and adoption of cyber insurance. Typically, it will mitigate some of your losses (incurred both by your own actions or inactions and those of others) and it will offset some of the consequential costs of security incidents such as the need to employ expert security professionals for a clean up after a data breach.

While cyber security insurance can be beneficial it is as well to check the small print to ascertain your responsibilities for keeping secure and to be certain that the insurer will pay out in the event of a breach.

They certainly have a role to play in your security defences but there is something of a Catch 22 in their implementation.

For your risk to be acceptable to your broker and insurer, they are going to want to see that you have reasonable security processes and associated controls in place. They will often mandate a standard, most often ISO 27001, against which you will be asked to demonstrate a good level of compliance. But that, in turn, means that you are then going to be much less likely to make any claim or that the value of such claim will be commensurately smaller. You may then judge that it is not worth paying the policy premium required.

If that is the case, you may well decide it is simpler to gain certification against ISO 27001 although this will not, admittedly, inoculate you from incidents and it isn't a cheap exercise. In which case, a blend of the two in return for reduced premiums, but also lower payouts, would be your best bet.

Protection through technology

In addition to simply using and operating digital technology right, defensive technology is all too often where security professionals start. This can range from simple anti-virus software to extensive (and expensive) Security Information and Event Management (SIEM) systems.

Buying shiny new technology is not only interesting for security professionals, but it also shows people that a job is being done. Of course, technology is an important part of the solution. But it needs to be proportionate (in terms both of expenditure and the effort to operate it) to the value of, and risks to, the assets it is defending.

It also needs to be used. In the Target data breach case (see the previous chapter), expensive technology had been purchased but largely disabled by operators who distrusted its effectiveness, and who ignored the warnings it delivered. Owning technology is not the same as using it.

Artificial intelligence is delivering a whole new set of technological solutions focussed around the detection of anomalous activity, such as files being accessed at unusual times and from unexpected places. These tools may well prove to be invaluable. But whatever tools are used it is important they are tested regularly to evaluate their continued effectiveness.

Role of the board in security

The importance of the board's involvement in, and ultimate accountability for, information security was noted right back in Chapter 1. And it cannot be overstated.

But, in as much as there is no single right answer for information security and it can be considerably affected by decisions on such matters as your organisation's appetite for risk, it is vital that they take the lead and then demonstrate that leadership.

It is the board's job to keep questioning themselves, their workforces, their business partners and other stakeholders on such matters as:

- How can you personally contribute to security?
- What are others doing that you could emulate?
- Is the assurance information you presently receive sufficient?
- What does the board not know that they don't know? (See the box "Unknown unknowns".)

UNKNOWN UNKNOWNS

In November 2017, *Discover* magazine carried a useful article about the importance, but also the difficulty, of seeking out the unknown unknowns (6).

> There are the unknown unknowns...They're an intellectual blind spot, and our brain loves to fill in blind spots. ... We are especially

prone to blind spots when it comes to physical phenomena or devices like earthquakes and bicycles, and complex systems like law or politics. The problem... is that most of us are not bicycle mechanics or political scientists, but we all have a passing familiarity with some of the surface features of bikes and politics. This smattering of vague knowledge can get us into trouble, because it takes a bit of research or expertise just to know how much you don't know about something. Without it, it can be hard to tell the difference between a deep pool of understanding and a shallow puddle.

Researching your lack of knowledge is surely well worth at least a small percentage of your annual security budget.

Having realistic expectations

Your governing body needs to become comfortable with the reality that 100% security is impossible. Security incidents will still occur. Near misses will happen even more often. Thus, a good level of resilience, as described in Chapter 13, is also important.

And there is a silver lining. Provided that you've implemented the right incident reporting and lesson learning processes, your board will be able to identify when it has experienced but survived a security incident. This experience can identify previously unrecognised risk tolerance, allowing you to update your risk appetite statements. In addition, any near misses may suggest to you controls that are working but could be working better: perhaps more training is needed or a small change to your processes. All this provides a contribution to your organisation's digital effectiveness and evidence of the effectiveness of your digital governance.

How are we doing?

It's important for board members to know how well their organisation is doing in regard to information security. In 2018, Marsh, a major provider of security insurance, funded and released a *Cyber and the City* report (7).

As well as other useful information and statistics, this report contained a maturity model in the form of a very straightforward self-assessment checklist for boards to use.

The statements that the board is asked to question are as follows:

- The main cyber threats for the firm have been identified and sized.
- There is an action plan to improve defence and response to these threats.
- Data assets are mapped and actions to secure them are clear.
- Supplier, customer, employee and infrastructure cyber risks are being managed.
- The plan includes independent testing against a recognised framework.
- The risk appetite statement provides control of cyber concentration risk.
- Insurance has been tested for its cyber coverage and counter-party risk.
- Preparations have been made to respond to a successful attack.
- Cyber insights are being shared and gained from peers.
- Regular board review material is provided to evidence the above.

Working with the Chief Information Security Officer

In organisations of any size, the requirement for a Chief Information Security Office (CISO) is critical. The role is complex, however, and it is unlikely that it is deliverable on a part-time basis.

The role should not be considered as a purely technical one and so you should not look for an appointee exclusively from the ranks of IT professionals. The CISO's focus should always be primarily on your business and they should have the ability to engage across your business as peers and with the board as credible as one of its strategic advisors.

CISO and/or DPO?

Organisations that wish to assure the privacy of employees and customers whose personal data is processed by their organisation may choose to appoint a Data Protection Officer (DPO) to oversee this. And in some cases, this will be mandatory: the EU's GDPR requires that DPOs are appointed by public authorities and organisations whose core activities involve *large scale, regular and systematic monitoring of individuals*. The role differs from that of the CISO. It is a monitoring and advisory role concerned with compliance with privacy regulation. In contrast, the CISO is responsible for delivering privacy as part of their responsibility to protect information. The requirement that the DPO should be independent from operations means that it might or might not be difficult to combine this role with that of the CISO, depending on how you structure their responsibilities.

Talking to the board about security

When engaging with your board, the CISO should:

- Use accessible and credible, jargon-free, simple and relevant language
- Visualise the topic to make it more engaging, making use of scenarios and parallels to bring the subject to life
- Outline the critical business issues (what stops us working, what the damage would be, are we safe, what action is needed) and avoid focussing on the technology issues
- Show trends rather than one-off statistics
- Explain what should be done and why but take account of the collateral manpower and resources likely to be required to operate the recommended controls
- Group technical issues like Wi-Fi access and network access into a small number of major business issues, e.g. customer experience or operational efficiency
- Keep written reports short and using proven tabular formats to aid reading, comprehension
- Focus on decisions the board can make.

Your CISO will of course need to provide you with evidence in the form of metrics and trends. There may be a temptation to showcase a large number of technical statistics such as the number of attempted and successful penetrations and the time taken to detect penetrations.

Most boards will be resolutely uninterested in such data. Instead, it is better to encourage the CISO to summarise data in the form of traffic lights outlining the people, process and technology layers we outlined earlier and focussing on strategic issues such as the rise of new threats and how they are being addressed, perhaps accompanied by some salutary tales of competitors who have failed to protect themselves.

However, your CISO can show you progress towards greater security with metrics such as:

- Fewer successful penetrations
- Increased compliance with standards
- Percentage of IT budget dedicated to security
- The number and nature of increases and decreases in risk
- The number of damaging events (over a particular threshold) causing data loss as calibrated by transactions affected
- The time taken to detect penetration
- The time taken to start responding
- The percentage of employees completing security training
- The percentage of procurement decisions that included cyber security consideration.
- A range of cultural measures, such as:
  - Perceptions of the importance of security
  - Specific acknowledgements of personal responsibility
  - Awareness of threats such as phishing emails
  - Willingness to go against cultural norms
  - Willingness to challenge strangers.

Getting help and contributing to increased professionalism

As discussed above, right back in Chapter 2 where we discussed community engagement, you should use what is already out there rather than create more of the same. But, at the same time, you really should be considering the next big things – like those in Chapter 14. However, community engagement is not purely a call on your altruism. You surely want to be on the front foot on the security issues related to these things, not just learning from others later on or, even worse, learning in arrears from the incidents you've had.

GOVERNMENT CYBER SUPPORT

In 2017, the UK Government formed the National Cyber Security Centre (NCSC) from a variety of predecessor component organisations. Its mission statement is: *Helping to make the UK the safest place to live and work online.*

Their website states that:

> We support the most critical organisations in the UK, the wider public sector, industry, SMEs as well as the general public. When incidents do occur, we provide effective incident response to minimise harm to the UK, help with recovery, and learn lessons for the future.

As part of that mission, the NCSC provides an increasingly wide range of materials to support organisations and particularly their boards. But the NCSC also encourages research and development of new techniques and updated guidance. For that, they really need you to get involved.

Delivering a culture of no regrets...

A board that can score itself highly, and evidence that scoring, in an honest assessment against any maturity model, need have no regrets about any security incidents that do still occur. They should be smaller in scale and shorter in duration than would otherwise be the case. And, even if there are major breaches, the board will still have done their duty. As we said earlier in this chapter, 100% security is impossible to guarantee.

References

1. ENISA. (2019). *NIS Directive*. [online] Available at: www.enisa.europa.eu/topics/nis-directive [Accessed 24 May 2019].
2. National Cyber Security Centre. (2019). *Cyber Essentials* [online]. Available at: www.cyberessentials.ncsc.gov.uk/ [Accessed 24 May 2019].
3. Australian Signals Directorate. (2019). *Essential Eight Explained* [online]. Available at: www.cyber.gov.au/publications/essential-eight-explained [Accessed 24 May 2019].

4. National Cyber Security Centre. (2018). *10 Steps to Cyber Security* [online]. Available at: www.ncsc.gov.uk/collection/10-steps-to-cyber-security [Accessed 24 May 2019].

5. Huntsman Security. (2019) *Benchmark Your Organisation's Cyber Security Posture* [online]. Available at: www.huntsmansecurity.com/products/security-scorecard/ [Accessed 24 May 2019].

6. Brotherton, B. (2015). *The Brain Has a Blind Spot for "Unknown Unknowns"* [online]. Available at: blogs.discovermagazine.com/crux/2015/11/17/brain-unknowns/ [Accessed 24 May 2019].

7. Marsh (2016). *Cyber and the City* [online]. Available at: www.marsh.com/content/dam/marsh/Documents/PDF/UK-en/Cyber%20and%20the%20city.pdf [Accessed 24 May 2019].

12

DELIVERING DIGITAL PRIVACY

Summary

Keeping personal data safe and private is an urgent priority for any organisation. This isn't just because of the requirements of the EU's General Data Privacy Regulation (GDPR), which has a global reach. Increasingly, there is a demand for privacy from ordinary consumers irritated by large technology companies taking them, and their data, for granted.

Privacy, therefore, needs to be at the forefront of the issues that organisational leaders address. That's because complying with the legal, ethical and social requirements of privacy is hard. In part this is because there are many misunderstandings about the rules – GDPR is still being interpreted by privacy practitioners, regulators and lawyers. In part, it is because privacy culture varies widely around the world. What is certain though is fines for non-compliance are high and the dangers of being found vicariously liable for mistakes made by employees are considerable. A robust structure for ensuring privacy is therefore essential.

Privacy is rapidly becoming one of the most contentious and emotive of digital governance topics with escalating public expectations. These are almost beyond what is deliverable by today's digital technologies and the uses being made of that technology.

Privacy has been made more complex by differences between what individuals in different countries see as personal data on the one hand and appropriate protection on the other. This is then affected by such incidents as the 2018 Cambridge Analytica scandal (1) and by what different national governments and supranational bodies like the EU see as appropriate protection. And it is further complicated by differences in what this legislation states is especially sensitive and requires additional, special, protection measures.

In this chapter, we address this growing challenge and position how it overlaps with a variety of, closely related, other considerations — such as risk management, security and resilience. And we want to emphasise that privacy is not a synonym for information security's focus on confidentiality even though a security breach causing disclosure of personal information is certainly also a privacy breach.

Privacy: an urgent priority

Privacy has become a more urgent priority for governing bodies since the EU's General Data Protection Regulation (GDPR) came into force across the EU, and with application globally to any organisation that processes the personal data of people visiting or living anywhere in the EU. (In the UK, the GDPR has been incorporated into the Data Protection Act 2018. For simplicity, in this chapter, we will still refer to it as GDPR.) That legislation will also be evolving further in the next few years, not least as case law develops precedents that need to be followed.

The 2016 Tesco Bank data breach (see text box on page 202) resulted in action by the financial services regulator, the Financial Conduct Authority (FCA), and not by the privacy regulator, the Information Commissioner's Office (ICO), but they were working together on the investigation. This shows that a privacy breach won't just attract action under data protection law, however.

**CYBER SECURITY AND PRIVACY:
WHY TESCO BANK WAS FINED BY THE FCA**

On 1 October 2018, the FCA fined Tesco Personal Finance plc (Tesco Bank) £16.4 million for its failures relating to a cyber-attack. Over a 48-hour period in November 2016, cyber criminals stole £2.26 million from more than 8,000 of Tesco Bank's 131,000 personal current account customers. The incident received significant press attention at the time. Nearly two years later the FCA delivered its verdict that Tesco Bank should have done more both before the attack and in response to it.

The fine was actually heavily discounted from £33,562,400, which the FCA stated that it would have imposed, were it not for a "mitigation credit" that reflected Tesco Bank's co-operation, remedial actions and a discount for their early settlement. Nonetheless, it sent a very strong message to the financial sector about the FCA's stance of no tolerance for banks that fail to protect customers from foreseeable risks. It was also intended to remind the financial markets to put in place proper systems and controls to mitigate the risk of an attack arising in the first place and to limit any harm to customers in the event of an attack.

The FCA had concluded that Tesco Bank had breached Principle 2 of the FCA's Principles for Businesses by failing to exercise due skill, care and diligence in the design and distribution of its debit cards; to configure specific authentication and fraud detection rules; to take appropriate action to prevent the foreseeable risk of fraud; and to respond to cyber breach with sufficient rigour, skill and urgency.

The FCA's Final Notice explained that Tesco Bank had been warned of the risk of attack in November 2015. Tesco Bank made changes to its credit cards to mitigate this risk. The degree to which its debit cards are still at risk is open to question, however.

Increasing volumes of increasingly intrusive data

Technology processes, such as data backup and use of cloud services that duplicate and replicate data, and new technologies, such as artificial intelligence and Big Data analytics that can create new personal data (e.g. about preferences and buying susceptibility), are making the tracking and management of personal data even harder. You need to know all the uses of the data, and the places where that takes place, if you are to ensure the privacy of personal information.

In April 2017, Seagate's *Data Age* 2025 (2) report forecast that, by that date, the volume of world data will have multiplied by a factor of 10 to 163 zettabytes and that 20% of that data would be critical to life. So perhaps you might extrapolate from this that at least 40% of it (64ZB) will be personal in nature.

But it is not, of course, merely a question of volume. A customer's credit card number is one thing, but if it is associated with their address, their date of birth, their past purchases and their email address, then this aggregation of data is undoubtedly very profound and justifies your governing body going to considerable lengths to ensure that it is secure. Better then not to store all such information together, even if that reduces the opportunities for data analytics.

GDPR: new rights and responsibilities

The EU's GDPR has the force of law in all of the EU's member states. The intention of GDPR is to deliver a much more level playing field across the member states than had previously applied. To do so, it codified a new set of rights and clarified a number of responsibilities.

New privacy rights

GDPR's new rights for individuals concerning their personal data are the general rights to:

- Be informed about what data is being collected, why it is being processed and how it is being processed.
- Have access to a copy of the data held about them.
- Obtain rectification of data that is incorrect or incomplete.
- Have personal data erased (however, this right only applies in certain limited circumstances).
- Restrict the processing of their data (again this is a limited right that does not always apply).
- Be given a copy of their data in a portable form so that they can reuse it elsewhere.

- Object to it being used for direct marketing.
- Not be subject to automated decision making including profiling.

There are some real challenges for organisations in delivering these rights, especially the right to erasure (often known as the right to be forgotten), as data is often replicated in all sorts of places such as different systems, back-ups and email. The implication of this is that you need to know all about your own data and, in particular, where it is to be found.

GDPR also sets new expectations in respect of how organisations will prepare for, handle and comply with the obligation to make mandatory reports of incidents, in an attempt to mitigate the scale of damage seen in recent privacy (and cyber) breaches, as shown in this chapter.

Increased penalties

Another aspect of privacy law that has changed through the adoption of the GDPR is that the fines for breaches of it have massively increased. It was only a few years ago that the maximum fine in the UK for a breach of the Data Protection Act 1998 was £500,000. This fine value had never been levied. Now, GDPR has introduced the concept of fines of up to 4% of annual global turnover or €20 million, whichever is greater.

Another change is that there is now a model for the calculation of a fine, based on a series of elements that address the full lifecycle of personal information processing. Under this approach, the UK Computing magazine reported that the fine levied upon Tesco (in the example on page 202) could have been as much as £1.9Bn if the breach had occurred under GDPR rather than the Data Protection Act 1998.

Large fines are now being levied when failures, breaches and abuses lead to losses of personal privacy. In 2019 British Airways was fined more than £183 million for the disclosure of 500,000 customers' details, including their credit card details, following a cyber breach. And, again in 2019, Facebook agreed with the US Federal Trade Commission (FTC) to pay a fine of $5 billion to resolve the data scandal triggered by Cambridge Analytica's harvesting of personal data from 270,000 people.

Not all GDPR infringements will, however, lead to fines. Besides the power to impose fines, the ICO now has a range of corrective sanctions to enforce the GDPR, including:

- Issuing warnings and reprimands
- Imposing a temporary or permanent ban on data processing
- Ordering the rectification, restriction or erasure of data
- Suspending data transfers to third countries.

These larger fines and other penalties are clearly intended to incentivise privacy by punishing transgressors on the one hand and sending a warning to your competitors on the other. Your concern should be the damage that incurring a penalty would cause to your reputation (and share price), and to your brands (and turnover) when this gets out. The damage is likely to be many times the value of the fine.

Governing bodies should therefore ensure that they can tell a confident story, with supporting evidence, of having carefully considered and agreed on the approach to each aspect of data protection. This then will significantly reduce any exposure to risks and limit any reputational damage in the wake of a data breach.

Other consequences of privacy breaches

Governing bodies will also need to be mindful of other considerations and consequences of privacy breaches. In the summer of 2017, the credit and identity referencing agency Equifax was the victim of a data breach that disclosed the information of more than 145 million Americans. Equifax's CIO and CISO both left the company. But the CIO exercised all of his stock options before the breach became public and the stock crashed, gaining him proceeds of nearly $1 million. He is now serving 4 months in prison, will spend a further year on supervised release and has to pay the profit made as a fine. In the meantime, Equifax's reputation has suffered further (3).

Privacy authorities

Most countries have a privacy regulator. In the USA, the Federal Trade Commission takes on this role. In China, it is the Ministry of Public Security. And in the UK, it is the ICO.

The ICO has committed itself to two key behavioural traits:

- To encourage good behaviours as much as to prosecute breaches, failures and poor practice
- To engage with organisations and individuals to optimise their preparations.

It makes sense therefore for UK organisations to ensure that a relationship is forged with the ICO and for organisations in other countries to identify their regulator and engage with it.

It's as well to note that the ICO interprets privacy law in one way and that regulators in other countries (such as the Dutch Data Protection Authority, the Autoriteit Persoonsgegevens) may well still interpret privacy in other ways. Where your operations extend to people outside one jurisdiction it's important for governance to take that into account.

Privacy compliance

Privacy oversight

Managing privacy should be a key consideration for any governing body and it would be good practice for it to include a privacy lead (perhaps also a suitably skilled Non-Executive Director to support and challenge the designated Executive Director Lead). You are likely to need to designate a digitally skilled Chief Privacy Officer (CPO). As we said earlier, there is considerable overlap between privacy, risk, security, audit and resilience. This suggests that some of these roles could be merged.

The roles will need adequate resources and an appropriate organisational framework through which to operate. But they do not need necessarily to be full-time roles. In Europe, Public Authorities and organisations undertaking large scale monitoring of individuals are required by the GDPR to have a Data Protection Officer (DPO). However, that doesn't mean they can't share one or employ one who also undertakes other roles. The right approach will depend on the amount of data protection work involved.

And whatever approach you take, you should certainly ensure that your CPO meets regularly with officers responsible for risk, security, audit and resilience to orchestrate their activities and thereby amplify their respective impacts. Your CPO, perhaps leading a team of Data Protection Officers, will then be accountable, and report regularly on progress and status, to your governing body.

Deciding on if, when and how to bring roles together can be a difficult process. Clearly, there is no one right answer for all, as demonstrated by the NHS case study in the text box below.

PRIVACY AND CONFUSION IN THE UK'S NHS

In the late 1980s and into the first half of the 1990s, the UK National Health Service (NHS) experienced a number of incidents related to patients, their data and their care, that gained significant negative media and public attention.

Dame Fiona Caldicott, at the time of writing the Chair of the Oxford University Hospitals NHS Trust, undertook formal enquiries into NHS data protection, information governance and privacy in 1996–7 and again in 2013 (4). This led to the creation of Patient Representatives – known more informally as Caldicott Guardians – alongside Data Protection Officers.

But the advent of GDPR seems to have been taken as an opportunity to complicate things once more. Many NHS Trusts now additionally have GDPR leads. How these roles interact and why they have not been brought together for cohesion and coherence is not immediately obvious. Things may well evolve as more is learned of GDPR's actual – rather than intended – consequences.

Processing data

The privacy rules are structured around the processing of personal data. Personal data, whether it is a computer file or a list of names written down on a sheet of paper, can be processed in a number of ways. In fact, pretty much anything you do to data counts as processing it: collecting it, storing it, analysing it, sorting it, sharing it, even deleting it. If you have personal data (and if you employ anyone, then it is certain that you do) you will be processing it. Some of that data is unlikely to be problematic. But other data is more sensitive.

Personal data

Under GDPR, personal data means any information relating to an identified or identifiable natural, living, person (known as the data subject). An identifiable natural person is one who can be identified, directly or indirectly, in particular by reference to an identifier such as a name, an identification number, location data, an online identifier or to one or more factors specific to the physical, physiological, genetic, mental, economic, cultural or social identity of that natural person.

There are then differences between your corporate and private persons. Your activities at work are not personal (subject to your rights and choice to use corporate digital technology for personal purposes, in which case you are loosening the controls on your own data) and nor, for example, is your email address at work. But your personal email (subject to any use you make of it for work purposes, in accordance with your organisation's policies) is another personal data item.

Surely though, it's all sensitive, right?

Is all personal data sensitive in the same way? The short answer is no. Personal information obviously exists in a number of domains and states. Treating it all in the same way would be onerous. Some data needs considerably more care than other data.

There is data that is in the public domain and has been put there in accordance with the law or because the individual data subject has put it out there. For example, in the UK if you are over the age of 18 your name and home address is likely to be found on the electoral roll which is publicly available, unless you have specifically asked that your details are not published. Similarly, the details of your car are supplied to many organisations. Details such as these are not considered to be of special significance and may well be used by organisations seeking to establish someone's identity or age.

In contrast, there is information that doesn't belong in the public domain: your internet browsing habits perhaps or the details of your monthly online supermarket shop. This is information that you are entitled to think of as rather more private.

And then there is sensitive data. GDPR defines a set of "special category data": this was called "sensitive personal data" under the UK's Data Protection Act 1998. Special category data is that which reveals a data subject's:

- Racial or ethnic origin
- Political opinions
- Religious and philosophical beliefs
- Trade union membership

- Genetic make up
- Biometric parameters including data for the purpose of uniquely iden-
tifying an individual
- Health
- Sex life and sexual orientation.

Processing this data is even more restricted than processing "ordinary" personal data.

Another set of data that is treated differently under the GDPR is children's data. In the latter case, you will need to be mindful of from whom you will get consent for your information storage and processing, however lawful that processing may be. For a 12-year-old, for example, you will need that consent from a parent or guardian, but a 16-year old is allowed to decide for themselves.

Lawful processing for specific purposes

However, it is not just about what personal data you collect, but what it is going to be used for. Your use of digital technology to process it must itself be for specifically defined purposes judged to be legal.

Processing must normally have a specific and lawful basis. If you can reasonably achieve your outcomes without including the personal data, you won't have a lawful basis. You must determine this before you begin processing, and you should document it in your policy and privacy notices. You can't decide to switch in mid-stream.

The six lawful bases are where:

- Consent has been given
- It's necessary to perform a contract with the data subject
- There is a legal obligation to do so
- It's for the vital safety or protection of individuals
- You are serving a public interest
- You have a legitimate interest to process the data.

Implicit versus de-facto consent

GDPR formalises a much more rigorous requirement for consent to obtained from individuals for processing. This explicit consent assumes that the individual has then been, unambiguously, informed of the purposes for which, and the manner in which, the information will be stored and used.

This was meant to see an end to consent by inaction, by merely ticking a box or requiring those individuals to read endless terms and conditions and other small print. It led to a plethora of mails into email inboxes notifying people of their online sign-ups and giving them the means to unsubscribe. But it is questionable whether this really meets the intention of the Regulation and associated UK legislation.

The intention was also, therefore, to prevent the capture of personal information for an original purpose and subsequent unilateral re-use of it for a new purpose at a later date.

There are inevitably also grey areas in respect of consent within the legislation that you should approach with care. For example, where an organisation has had an individual on its mailing list for a very long time, originally at the individual's request (though the evidence of that may have been lost), has been sending them promotional mailings without receiving any objections and perhaps even had take-up of the service, is this consent adequate? It clearly can be asserted to be low risk to the individual – especially if that individual whilst not wanting the service or mailings has perhaps re-routed the notifications to their email junk file – but it is not explicit as such. Only time will really tell.

Privacy: a trade-off between principle and pragmatism

In 2018, a number of energy suppliers went into administration. No doubt with the entirely honourable intention of ensuring continued energy supply, OFGEM, as the UK Energy Regulator, arranged unilateral transfer of the affected customers' accounts (and thus their personal details) to alternative suppliers. But OFGEM did so without getting *any* consent from those customers to do so.

At least one customer, known to the authors, had previously been with one of these energy suppliers and had already switched away from them for a better deal. What's more, the customer had been obliged to take out a successful court action for the recovery of very significant over-charging by that energy provider.

Was OFGEM right in passing on their details to a new supplier? Had they not done so the customer might have experienced energy-supply problems.

But even so, OFGEM's actions could be said to have damaged the privacy rights of that customer.

Governing bodies need to know that personal data is processed lawfully and that records such as consents are being kept. After all, it will be important to have evidence that your processing is "fair" if it comes to any complaint. This has a resource implication for the organisation, but the task is a critically important one that the governing body must therefore be prepared to resource.

Subject Access Requests

Knowing what information you hold on individuals and how it is processed is important for several reasons we have already covered. But another reason is to be able to handle, efficiently, effectively and accurately, data subject's requests to obtain details of the data that you hold about them. These are known as Subject Access Requests (SARs).

Clearly, you need to take steps to define fully your processes for these requests, the individuals who will respond to them for you, with whom they will consult and from whom they will obtain approval of the proposed responses. Whilst this is only good sense, you need to ensure it works smoothly to ensure you don't breach the (reduced) time limitations that now apply and incur the penalties that can be levied for doing so.

Getting it right
Questions of ethics and policy

Establishing your organisation's chosen positions on the privacy of personal information ought to take account of the expectations of your stakeholders including regulators.

What you will clearly be trying to get to a position on is the balances to be struck between legal requirements and what is reasonable, what is ethical and considerate and what is lawful, what is in the spirit of the legislation and what could not be said to be.

Here, we are taking it as a given that you will instruct and train your personnel to behave responsibly and cautiously with personal data. Three commonly adopted principles in this respect that you would do well to reflect in the guidelines on privacy that are set out for your employees are:

- Treat others as you would want to be treated yourself.
- Do not share data without specific approval and share in approved manners only.
- If it does not feel right, don't do it.

This won't guarantee that you will always get it right, but it is a good departure point.

LAWFUL, APPROPRIATE AND PRACTICAL

The UK Scout Association has a framework of medal awards for its adult leaders through which appreciation is shown for their many years of voluntary service. Once an award has been agreed at UK level, lists of recipients are shared with each county and the counties tend to break down that list into their districts and thence to the Scout Groups.

In 2019, at least one county and district both placed their lists on their Facebook sites and one group was planning to produce a press announcement about one of their leader's awards for 50 years of service. But the leader concerned had not been consulted about any of the data processing involved.

The Facebook posts were to closed communities and the posts were a list of (only) names and medal awarded; and all this was done with the best of intentions. But was it reasonable or considerate? The process was too lax. A quick email to the recipient would have established whether they had any concerns about this publicity: simple courtesy really.

While it is important to ensure that organisations are compliant with privacy rules, it's also important to avoid excessive responses. For example, a large British environmental charity keeps records of the volunteers attending working sessions. Although the charity has names and contact details for next of kin, in case of accident or injuries, it has taken to not pre-including them in signing-in sheets.

This was driven by fear of information disclosure to other volunteers. However, the circumstances prevailing when the sheets are filled in at the start of the session are such that there is no opportunity for any third party, when filling in their own details, to copy others' details for subsequent misuse.

A small matter perhaps but certainly a waste of precious volunteer time – when those volunteers had no such concerns before.

Cultural dilemmas, limitations and exclusions

Whilst GDPR is driving a strong convergence of good practice across Europe, there are more marked differences between the European view and others around the world, notably the USA, Russia and China, to name just three.

In the light of GDPR, a new EU-US Privacy Shield framework replaced a previous Safe Harbour framework. It is a binding legal instrument under European law which can be used as a legal basis for transferring personal data to the USA. On 12 July 2016, the European Commission issued its formal adequacy decision on the Privacy Shield.

The creation of this Shield arrangement has, in turn, ushered in a European Commission process for establishing the adequacy of other countries' data protection arrangements. As of March 2019, personal data can flow from the EU (and Norway, Liechtenstein and Iceland) to Andorra, Argentina, Canada (commercial organisations), Faroe Islands, Guernsey, Israel, Isle of Man, Japan, Jersey, New Zealand, Switzerland and Uruguay without any further safeguards being necessary.

All of this helps international trade and organisations that operate across borders. But it is then vital that you know the law and the reality of data protection in all of the countries that you operate and trade in. Operational efficiency suggests that you will need to establish a common, perhaps baseline, approach to data protection across your organisation that is then topped up with local additions when and if that baseline is not sufficient.

But that still raises the question of the adequacy of data protection arrangements across your supply chains. It is true that many US companies, for example, including Facebook, Apple and Google, have established, or are in the process of building, EU-based data centres to handle data for EU citizens. Google, for instance, lists four data centres within Europe, including one in Ireland. But the adequacy of these is still emerging and is being complicated by issues such as the use of cloud services and requirements for 24 × 7 availability.

In Chapter 2, we discussed the ethics of Facebook enrolling teens in a scheme to collect extensive online behaviour and other data in return for $20. But what about the security of that personal data if the scheme extends beyond US shores?

Other required processes

In this chapter, we have stressed the importance of a structured approach to privacy, confidentiality and data protection. And that this should cover the whole processing lifecycle, following a "privacy by design and by practice" approach, akin to that for security that we discussed in Chapter 11.

Similarly, we have recognised the overlaps between privacy and risk, security and resilience. The best way in which to ensure that potential impacts caused by privacy breaches are assessed, and properly documented for governing body approval, is to ensure that your risk assessment processes include the requisite consideration of privacy, whilst they themselves exploit a true and current record of your information assets.

The results of these assessments will then be a major input to the organisation's Privacy Policy and Practices which the governing body should sign off.

The logic chain will then make it relatively straightforward for your CPO to draft the Privacy Impact Notices that should be distributed to all staff and all data subjects, as part of the responsibility you have to inform them of what processing is happening and why and what they should do to demonstrate their acceptance, i.e. give their explicit consent, or otherwise.

Then you will be well placed to provide structured training and awareness, covering both the principles of data protection and the practices you want applied. And then, so long as you have addressed the other aspects of data protection discussed above, you should be in a good place to process personal data safely, securely and with a low risk of sanction from regulators.

Employers' vicarious liability

There have been many instances of claims that employers should not be held liable, even vicariously, for the misbehaviours of their employees in respect of them misusing personal information. Subject to an awaited Supreme Court ruling, it seems that a case involving Morrisons is set to conclude that employers are vicariously liable in such circumstances.

GROCERY CHAIN GUILTY AFTER AN EMPLOYEE STEALS DATA

In 2014, Andrew Skelton, who worked at grocery retailer Morrisons (5), leaked the payroll data of nearly 100,000 staff. Skelton, who worked as a senior internal auditor at the Bradford head office, appeared to have a grudge over a previous disciplinary incident. He was arrested and convicted in July 2015, at Bradford Crown Court, for breaches of the Computer Misuse Act 1980 and the Data Protection Act 1998 and sentenced to 8 years imprisonment.

The next year, in the first data leak class action in the UK, more than 5,000 Morrisons employees brought a claim against the company for damages caused by the privacy breach. After the initial set of proceedings, the UK's Appeal Court had to decide first whether Morrisons were directly liable for the breach under the Data Protection Act 1998 and second whether they would be vicariously liable for the malicious acts of their employee.

On the first point, the Court found that Morrisons was not the relevant controller. Skelton was directly liable under the Data Protection Act 1998. In addition, it was found that Morrisons had done all they could to secure the data with appropriate technical and organisational measures. As a result, the Information Commissioner's Office (the privacy regulator in the UK) did not pursue Morrisons for the data breach.

But should Morrisons be vicariously liable for Skelton's malicious acts when his intentions were to cause maximum damage to Morrisons? The answer the Appeal Court gave was "yes". In this case, vicarious liability applies regardless of intention and consequences.

At the time of writing the case has yet to go to the Supreme Court. However, vicarious liability has been demonstrated in a number of areas and employers always need to be aware of this danger. Employees who have access to large amounts of personal data are a potential liability for business. The answers are good staff training, robust processes that stop or at least deter data theft and perhaps cyber insurance as it is very likely that in the event of a breach you *will* be vicariously liable.

All governing bodies need to ensure they roll out a program of compulsory training to prevent primary liability for any data breaches of their employees, whether accidental or malicious, and to ensure their insurer (if they have one) will indemnify them. Insurers are unlikely to indemnify you if you have not had any compulsory and properly tracked training in place.

But why should that training be compulsory? The biggest threat to any employer for data leaks is not a malicious employee but more commonly a simple human error. In addition to internal policies and procedures, training in them is paramount.

Some data breaches now require mandatory reporting to the ICO. When doing so, one of the questions posed is whether the person involved in the breach has had data protection training within the last 2 years. An insurer may well escape liability for a bigger claim if it can be demonstrated that the person involved has had appropriate data protection training. The cost of sanctions could potentially far exceed any investment your organisation makes in such training.

Embedding privacy and minimising risk

But there will always be other measures that you could choose to take, on a case-by-case basis, to reduce your privacy risks further.

Whilst you will always need the ability to process data, it won't always be necessary to process *personal* data. If you don't need to use personal data then don't keep it. If it can be anonymised by deleting the part of the data (like names) that identifies people and still meet your purpose then do that.

Anonymisation means that data can never be converted back into personal data. If you feel that you may, at some point in the future, need to tie the data back to living individuals, consider using pseudonymisation techniques where the names are replaced with pseudonyms in one file and where the names together with the pseudonyms but without the rest of the data are kept in another file. Just remember not to keep these two files together in the same place!

Encrypting data is another technique. This means that no one without a key should be able to look at your data. As the costs of digital encryption technologies fall, and user familiarity with them rises, it is likely to become more and more practical to encrypt data more often when you are sharing it (for instance, over email) as when it is at rest in a database. That said, contrary to what encryption salesmen would have you believe, it isn't a total solution for privacy by any means.

A logically followed, documented and rigorously implemented process, in which your people have been appropriately trained, will normally facilitate your personal data processing and should see you right if your Privacy Regulator comes knocking.

References

1. Wong, J.C. (2019). *Facebook Acknowledges Concerns over Cambridge Analytica Emerged Earlier Than Reported* [online]. Available at: www.theguardian.com/uk-news/2019/mar/21/facebook-knew-of-cambridge-analytica-data-misuse-earlier-than-reported-court-filing [Accessed 24 May 2019].

2. Seagate. (2019). *Data Age 2025* [online]. Available at: www.seagate.com/gb/en/our-story/data-age-2025/ [Accessed 24 May 2019].

3. www.finextra.com/newsarticle/34059/former-equifax-cio-jailed-for-insider-trading [Accessed 6 August 2019]

4. Dame Fiona Caldicott Is the UK's National Data Guardian: Gov.UK. (2019) *National Data Guardian. Dame Fiona Caldicot* [online]. Available at: www.gov.uk/government/people/fiona-caldicott [Accessed 24 May 2019].

5. BBC News. (2018). *Morrisons Loses Data Leak Challenge.* [online] Available at: www.bbc.co.uk/news/business-45943735 [Accessed 24 May 2019].

13

THINK DIGITAL RESILIENCE

Summary

There's no such thing as 100% security. Accidents happen and damage occurs. This is as true for digital technology as for anything else. Cyber breaches can happen. Critical systems can fail. Reputations can be damaged online. Resilience is about enabling organisations to predict and respond to this damage and indeed to move from surviving to thriving. In this, resilience is not just another name for business continuity; it goes further and deeper.

And because accidents happen, resilience is essential. It takes effort to make an organisation resilient. First, you need to understand the risks you face, how they can be minimised, and how best to respond when incidents happen. Plans need to be drawn up and they need to be tested, partly to see if they work, partly to give people the experience of handling a crisis. Stress must be managed. Communication handled positively, internally and externally. And when the worst happens, reviews must be undertaken so that learnings can be found. None of this happens naturally. It takes leaders who know they are accountable and who have a vision of what good resilience looks like.

The velocity of change in the world today and the host of hostile actions going on, on a daily basis, have served to bring resilience to the fore as a major challenge for governing bodies. It is galling, as well as damaging and disruptive, to be collateral damage in events that otherwise had nothing to do with you.

Meanwhile, the world is very reliant on digital technology. This is particularly true in the UK where one eighth of GDP comes from the digital economy – the highest in the G20. But all developed economies are highly dependent. Digital organisations are, by definition, intrinsically reliant on their digital machinery. This has been true for many years. In 1994, a well-known UK advertising agency found itself entirely unable to trade when all the memory chips were removed from its PCs as part of a sophisticated burglary.

Your organisation, in contrast, will already have developed considerable intrinsic resilience if you have applied good digital governance. But resilience is very much an attitude of mind that needs to be cultured and continuously applied. And it needs to interact with other areas of focus, risk management, security and privacy in particular.

A grandiose name for business continuity?

Resilience is not just another name for business continuity. It needs to work with business continuity and indeed with disaster recovery, contingency planning and crisis management. But it goes both further and deeper, as well as being more focussed on preventions than on cures.

Resilience: a few definitions

Resilience. The ability of an organisation to anticipate, prepare for, withstand, respond to and adapt to both incremental change and sudden disruptions in order to survive and prosper.

Disaster recovery. Plans and facilities for the reconstruction – or at best the re-provisioning – of services, typically of digital technology, after a fundamental failure or loss.

Contingency planning. Alternative operational arrangements, that may or may not include the invocation of disaster recovery, where the incident goes beyond digital technology, e.g. into operational areas and/or other facilities of the organisation.

Business continuity strategy and planning. A broader-based solution, including preventive elements, for a more comprehensive response to more significant events.

Crisis management planning. Arrangements for response, actively led by the governing body, to manage events of such magnitude that they could affect the long-term viability of the organisation (Figure 13.1).

The scope of resilience

Organisational resilience (and within that, therefore, digital resilience) is the ability to prevent the occurrence of incidents, to detect incidents that have occurred (or will occur) and to reduce their impact upon the organisation. Its emphasis is upon sustaining and recovering operation effectively and efficiently as a springboard to repairing, recovering and ideally better-configuring those operations.

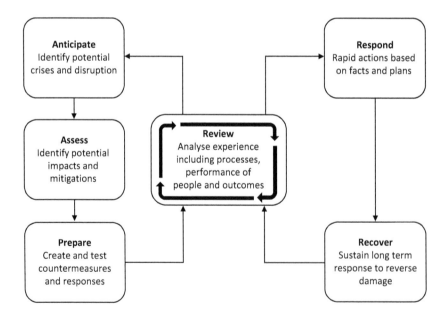

Figure 13.1 A simplified resilience management framework.

Resilience is a multi-faceted activity that also applies to all the different elements of your operations but especially to your digital operations. It is by no means then simply a technology issue either. Rather it is an inclusive process. It concerns your staff, their resilience to stress, e.g. when adapting and using new digital technologies and services, their resilience in respect of the roles they perform for you, and their roles in respect of contributing to resilience.

It is a challenge to identify what is *not* involved. Everything technological, physical, procedural and human is in scope. It is also inter-organisational. For instance, it may involve an organisation's supply chain. One example of this is a popular drinks maker which only has 72 hours stock of drink cans and is critically reliant on its can-maker. But the can-maker has only one factory serving the whole of Europe. This does not appear to be a very resilient supply chain and the drinks company could be severely affected if their supplier's highly automated plant was hit by malware.

There are also real benefits to taking pan-sector and cross-sector approaches wherever possible, e.g. to ensure that a sole supplier, depended upon by all in a sector, won't suffer supply problems. And it is at least as relevant to the public sector as it is to the private sector.

A simple example of digital resilience is the succession planning that you require from your HR function. Who could maintain the website if your webmaster is ill? Who could continue with the development of your new AI diagnostic engine if the head of data science left? It's about facilities too. Is there spare capacity on your data connections if you are hit with a "denial of service" (DDoS) cyber attack? Is your data backed up and available if you are hit by a ransomware attack?

Foundations for resilience

The specific solutions to the digital resilience needs of your organisation will be largely unique to it. However, there are some fundamentals that are vital for success:

- **Governance and accountability**. The governing body are ultimately accountable for ensuring that the organisation has an appropriate level of resilience. It cannot be passed off to anywhere else in the organisation, such as the IT department, without then incurring both unwanted and insufficient results.

- **Leadership and culture**. The organisation's leadership, not least by its own example, will build resilience thinking into the organisation's culture and behaviours so that it becomes as automatic as it is possible to make it.
- **Common vision and goals**. The governing body must clearly communicate their vision of how digital technology will support the organisation's purpose, strategy, goals and objectives if actions then taken by others are to be correctly aligned.

Situational awareness and responding to the risks

In Chapter 2, we discussed the need for your organisation to understand the nature and extent of the risks caused by technology that it faces. This, in no small degree, will be derived from your situational awareness, your knowledge of what is happening both in the external and internal context within which it operates. We also discussed the need for you to be clear about what appetite the organisation has for carrying those risks, and the need to put controls in place to mitigate those amounts of risk for which there isn't the appetite or ability to tolerate. This understanding needs to be your primary driver in deriving the organisation's resilience strategy from which plans can subsequently be developed by others.

Thinking outcomes, not triggers

But that strategy needs to be developed in respect of anticipated outcomes. Whilst it is important to try to anticipate the potential triggers – to help you focus on what controls might be appropriate – your organisation needs to end up with a manageable set of plans and not shelves full of documents that are 90% the same. For example, your chosen priority in your strategy could be to protect reputation in the event of a major systems failure (whatever might have been the trigger event). Your organisation will therefore need both plans for restoring the system and for communicating with stakeholders. The added benefit of these plans is that they should still be usable for trigger events that you had not thought of too. For instance, your communications plans will be largely effective in protecting reputation irrespective of the trigger event, although the plans for restoring service levels might have to be amended if an unexpected type of failure happened. But even then, it is likely to be the detailed technology-related actions that change, which would rarely be held in the recovery plan itself.

Devolved but co-ordinated resilience planning

The number of organisations that have a Corporate Resilience Officer (CReO) is still low but it is rising fast. In some cases, the Corporate Risk Officer (CRO) or Corporate Information Security Officer (CISO) has been upcycled into the role (and in some cases they've even had some more money for the considerable additional responsibilities!).

But everyone has a role to play. Therefore, it is vital that your CReO is treated by you as a facilitator and co-ordinator and not the principal deliverer of resilience. You will therefore need to define the responsibilities that you are delegating to them and to your other managers.

Balancing tactical decisions with strategic goals

It is important that you put limits on the authorities that you give them. This is because local and tactical responses, when selected without appropriate oversight from your governing body, e.g. a panicked response to a problem with AI perhaps or any form of over-reaction to an interruption of a minor system, can undermine or conflict with the organisation's defined strategic objectives, especially if these have not been disseminated adequately across the organisation.

A capability to adapt

All organisations tend towards a consolidation on the *status quo*, and in response to incidents that impact the organisation's continuity, hanker for the *status quo ante*. In leading your organisation, you should retain a focus on building up its ability to adapt, to reach for better outcomes, not just the same as before. You should seek such capabilities as:

- **Resourceful action:** Reallocating assets and other resources to address issues more effectively, for instance, by using cloud computing solutions rather than separate locally based servers to store data.

- **Promoting innovation:** Finding new ways of doing the old things, for instance, using additive manufacturing to create prototype spare parts rather than continuing to wrangle metal in a workshop or be dependent upon a fragile supply chain.
- **Encouraging flexibility and agility:** Doing things more quickly – perhaps using file-sharing technology to enable disparate teams to work together more easily.
- **Sharing knowledge** of errors, incidents, failures, good practices, effective solutions and successes, for example, encouraging people to share their experience using video, that is then easily accessed and securely stored rather than forcing them to write laborious reports that are then forgotten about and lost in a filing cabinet.
- **Seeking knowledge**, especially from outside the organisation and from the diagnosis and extrapolation of near misses, e.g. arranging a private session at your next sector conference to discuss any recent events and solutions for them.

Avoid over-dependence

Dependence upon a service provider can often occur insidiously as other providers are dropped, more operations are outsourced, or uploaded to the Cloud or when more and more services are taken up from the same supplier. The latter may seem attractive but it might not be very resilient. We aren't sure what to make of Facebook's intention to provide a digital currency, but could it make it difficult in the future for your organisation to robustly manage the services that they provide to you?

Resilience challenges

The adoption of digital technology can often lead to single points of failure if the systems design has not been subject to resilience consideration. Increased internet thinking, e.g. thinking in terms of data duplication and secure access to it over the internet rather than across a private network, would arguably address this at a stroke.

The complexity of digital technology infrastructures can be a very real bar to truly resilient services. For example, in 2017, British Airways, which like most airlines has a fixation on the availability of its systems and thus a major focus on digital resilience, suffered a catastrophic power outage. This disrupted the travel plans of tens of thousands of people and plunged its hubs at Heathrow and Gatwick into chaos with around 800 flights having to be cancelled.

It was subsequently reported that the outage had been caused by a staff blunder. A power supply unit at one of BA's many data centres, which was in perfect working order, was shut down, triggering an uncontrolled rebooting of the data centre's system and a power surge which in turn damaged equipment and caused the wider systems outage. Initial investigation by BA claimed that this was a human error by a contractor rather than a deliberate act, although the contractor denied this.

The International Airlines Group (IAG), BA's parent, considered it necessary to send an explanatory email to all staff and BA's Chief Executive spent many uncomfortable hours being challenged by the media. Wherever the fault lay, clearly, BA's resilience planning had overlooked a key potential failure scenario. The assumption perhaps was that the data centre was supplied with an "uninterruptable power supply" (UPS, a supply with a battery backup plus a generator backup) and so there was no need to explore the scenario of the data centre's power supply being interrupted. Or perhaps the fact that the UPS had been installed decades previously simply meant that no one thought about it.

Testing and rehearsal

The Prussian Field Marshall Helmuth von Moltke is well known for saying "no plan ever survives contact with the enemy". Boxer Mike Tyson put it more robustly: "Everybody has a plan until they get punched in the mouth".

But that's not to say that plans are useless. They may well have many sound elements on which the ability to adapt can be anchored. However, a documented plan is also only as good as the people that execute it.

It is therefore vital that you test, rehearse and stress-test your plans. This allows you to confirm the viability of the plan, look for snags or details that have been omitted and familiarise you and your staff with the actions that should be taken.

And with digital technology these tests are essential. Everything is connected (as BA found out); things happen very quickly (because communication between machines and systems can be near instantaneous); and the often *ad hoc* nature of technological implementation and development mean that the precise structure and nature of digital systems is often undocumented and little understood. Without testing, you can have little idea of whether your plans will work.

As well as testing your plans, it is also possible to go further. When you have confidence in your basic plans consider throwing in curveballs, scenario complications and twists that stress-test your capabilities. For instance, preparing for a busy shopping period such as Black Friday or the release of tickets for a new event could well warrant first stress-testing the ability of ecommerce websites (and associated systems such as call centres) to cope with massively increased loads, checking the capacities of your fall-back processing systems and teams if they should still fail and ensuring that your skeleton communications briefs are suited to these failures.

Similarly, a digital twin (see Glossary) could be built that replicates a complex system such as the infrastructure of a city. The whole system can then be tested against various stresses. In the UK, the city of Newcastle (1) has used technology to recreate the whole city digitally with the aim of helping planners to stress-test the infrastructures in response to changes in the population size, pollution and even climate change.

Keeping records of your tests is good practice but also enables you to demonstrate the thoroughness of your preparations, helping you to prove the compliance of your processes, as described in Chapter 10.

Any testing must include involving the governing body so that its members are familiarised with their roles and the limits, both natural and created, of those roles.

Managing stress

Having mentioned stress, it is important to recognise that crises are very high-stress situations. And this is especially true with digital technology. Take the example of a social media crisis. Perhaps your organisation has developed a faulty product that has put children at risk and you are experiencing a considerable social media backlash. Such a situation will be very stressful as well as difficult to deal with:

- Members of the public in large numbers with be talking about your organisation, and probably about you personally. They are unlikely to stint with their criticisms or moderate their language. This can be very hurtful and distracting.

- Activists may discover where your family lives or where your children go to school. They may target them as well as a way of targeting you. This naturally is very upsetting.
- The press will get hold of the story and threaten to publish it, making the story's reach and the reputational damage far worse.
- Your employees will be stressed by the unaccustomed attention, by the call volumes to your help-lines, by the volume of mails in their email inboxes, by their managers' repeated enquiries about status. Some may simply leave and go home, leaving you short-handed. Others' productivity will understandably decline.
- Stakeholders such as bankers, suppliers and shareholders will be in touch; some will be worried and others angry, adding to your stress levels.
- So much is happening, so quickly, with the situation changing by the second and priorities constantly shifting...

Testing a situation like this by running a realistic scenario may well show that an otherwise great leader folds up under stress or, perhaps worse, develops a Superman complex. This person is highly valuable in more common situations. So how can they be supported?

And of course, they may not even be there. Perhaps they are on holiday, or ill, or are on an aeroplane and uncontactable. Knowing how well their deputies cope (and indeed knowing who their deputies would be) is valuable learning.

As well as you then needing to find an alternate person to undertake the role, you should take careful measures about that individual. After all, they have not failed and are entitled to fair treatment.

Stress can be a major problem, both during and after a crisis. In any significant crisis then, it is best that arrangements are in place for people to obtain counselling and support if they are upset.

Importance of effective communications

Communications always need to be effective, of course, internally but also externally in these days of social media storms such as the one we have just described.

There is no substitute for effective communication by accountable individuals. But this doesn't have to be improvised. Many potential crises are predictable to a degree and for them at least some communications can, and wherever possible should, be pre-scripted. It can often be the case that even if you are successfully dealing with a crisis, if you don't appear to be as well then in fact, you are failing.

Good communication requires training. Knowing how to deal with the media, and through them the public, takes skill. And a different set of skills is needed for dealing with the public directly on social media or your website. Regulators, bankers and shareholders may need yet a different approach. And of course, you will also need to ensure that your employees are feeling confident and supported.

TALKTALK'S CEO AND SEQUENTIAL ATTACKS

In 2015, the UK telecommunications company TalkTalk was the target for a cyberattack on a website they controlled that was essentially left over from a predecessor business, Tiscali. This website contained out-of-date database software. As a result, some of the website's pages were vulnerable to a well-known hacking technique called "SQL injection". Hackers got into the database and stole the personal data of a large number of customers.

Once it was clear that a breach had happened, TalkTalk took the bold, and ideal but as yet still unusual, step of engaging with the public in a very overt way. It did so with the Chief Executive fronting up interviews with the media to explain what was happening and to reassure the public.

An admirable strategy perhaps, designed to show the company was taking the breach very seriously. Unfortunately, the CEO was not an expert in cyber security and had not been well briefed in this fairly complex subject. She also experienced gender-based criticism that was unfounded and nearly illegal but certainly disruptive. She made a number of hasty (and as they turned out incorrect) statements about the size of the breach as well as talking about a "sequential attack" instead of an SQL injection attack. As the attack unfolded over the course of the night, she bravely stayed on duty, although clearly exhausted.

Was this a good strategy? Having the CEO there showed that the company was taking the attack seriously and being upfront was the socially responsible thing to do. However, the CEO was unable to answer technical questions sufficiently well: being able to delegate to an expert on her team would perhaps have been sensible. In addition, she shared speculative details about the number of files stolen which exaggerated the size of the breach. It is often hard to pin down what is actually going on in a cyber security attack (hackers may be using several techniques at the same times in an attempt to destabilise defence teams) and how many files/customers have been affected, so keeping strictly to the known facts is generally sensible.

The best conclusion to reach is that TalkTalk did the right thing but in the wrong way, perhaps because they had underestimated the complexity of digital technology. We will leave them with the last word, however: "The TalkTalk attack was notable for our decision to be open and honest with our customers from the outset. This gave them the best chance of protecting themselves".

Lesson learning and review

Throughout your resilience activities, both pro-active and reactive, you should identify lessons to learn and changes to make. You should ensure that you have processes in place to capture and address those lessons. As in the other fields of digital governance, you also ought to put in place arrangements for periodic reviews of your resilience arrangements, again following the three lines of defence model described previously in Chapter 10.

Reference

1. White, T. (2019). *Newcastle's 'Digital Twin' to Help City Plan for Disasters.* [online] The Guardian. Available at: www.theguardian.com/cities/2018/dec/30/newcastles-digital-twin-to-help-city-plan-for-disasters [Accessed 24 May 2019].

14

EMERGING DIGITAL TECHNOLOGIES

Summary

Digital technology is constantly changing and that's one of the reasons it is so difficult to manage. No one can know what is around the corner. Perhaps a commercially available quantum computer will be launched next week that somehow renders existing cyber security null and void. But organisations can protect themselves to a degree by monitoring technology developments and identifying, and following, those that are most likely to affect them.

In this final chapter, we review the nature of a few emerging technologies which are starting to disrupt a number of industries: Big Data analysis; robotic process automation; artificial intelligence; wearable computing; the Internet of Things; 3D printing; augmented reality; blockchain; and brain-computer interfaces. Any and all of these may well turn out to create massive and unexpected changes to your industry. Keep a watchful eye on them and any other technologies that start to look significant.

It is obvious that digital technology is in a constant state of flux and development. But your adoption of it raises considerable new digital governance questions to which you will need to find answers. And if you need to develop totally new solutions, you may well be taking yourself into uncharted waters. This is not always a great place to be, as we discussed in Chapter 3.

Much technology offers the prospect of freeing your people from the 3Ds – Dirty, Dangerous and Dull work, in favour of higher value and more rewarding tasks. It can also open up opportunities for more efficient and more effective delivery. But for it to do so, you will need to be sure that you know all there is (at the time) to know!

Whilst this chapter, alone, cannot hope to achieve that, let's take a look at a number of the most significant emerging technologies during 2019. It is worth remembering that, as in almost all technology changes, the names and terms get used differently, by different people, for their own ends, often in a desperate attempt to get the benefits of that technology assigned to them too, merely by association.

Big Data

Big Data has been referred to by many as a phenomenon. It often underpins other emerging technologies, providing both the place that inputs to new processes are obtained from and the place that outputs from them are deposited. But Big Data itself also provides the opportunity to gain richer information from that mass of data.

In a number of places elsewhere in this book, we have looked at the challenges related to having large amounts of data on hand – typically made up of different types of files and records – and the analytics that can then be performed upon them.

But Big Data is not just about the ever-increasing amount of storage we all want on our machines to store the massively flabby file types and sizes that have emerged as a result of the costs of that storage having massively diminished. It's far more than that. It is about quantum leaps in that data: petabytes, exabytes and zettabytes of data rather than the megabytes, gigabytes and terabytes of data which you can already store on your home PC (Table 14.1). The bytes are certainly getting bigger!

Big Data is really a catchphrase used to describe a massive volume of both structured and unstructured data. Structured data is data that has been

Table 14.1

Byte	1	A single character
Kilobyte	1,000	A page of text
Megabyte	1,000,000	A long book or a photograph
Gigabyte	1,000,000,000	7 minutes of HD TV
Terabyte	1,000,000,000,000	50,000 trees made into paper and printed
Petabyte	1,000,000,000,000,000	13 years of HD TV
Exabyte	1,000,000,000,000,000,000	250 million DVDs
Zettabyte	1,000,000,000,000,000,000,000	3% of the world's data storage in 2018
Yottabyte	1,000,000,000,000,000,000,000,000	A byte for every star in the universe

organised into a pre-designated form, typically of rows and columns that can be easily searched in a database. Unstructured data is data that hasn't got a formal structure, such as emails, market research reports and websites.

It is usually so large that in all but the largest enterprises it is too difficult to process using traditional software and run on the mill processing platforms. And this is because it has three characteristics, the 3Vs – the volume of data, the variety of formats and the velocity of processing that is required. If you want to process Big Data you are thus likely to need to use specialist third parties for the processing capabilities you require, thereby adding a new complication to your operations.

Big Data can be the result of capturing the total collections of any given data type, e.g. all the attempted configurations of potential vehicles created by users in car manufacturers' configuration tools provided on their public websites, and/or the increased segmentation of overall processes into the discrete steps taken.

Using Big Data

Health care services are starting to use Big Data to reduce the cost of treatment or increase the success of treatment by pulling together and analysing a holistic view of the patient's health, identifying connections that may not have been obvious to practitioners restricted to paper notes in different departments, NHS trusts and Care Group silos.

Cyber security professionals are using Big Data to pull information together from different sources – the way employees are handling files or behaving at work, the information contained in files, the types of hardware and software being used, network activity and activity on the Dark Web. Together this data can provide very valuable clues about where vulnerabilities lie and where attacks can be expected.

The entertainment industry uses professional reviews, purchasing patterns, news reports, social media comments and demographic data to provide recommendations to their users as well as to help them develop new products.

The attraction of Big Data is the richness, accuracy and increased value of the results that can be derived from it.

But there are real costs to it, over and above the costs of data storage and platforms to process it on. As it is a mixture of structured and unstructured data, you will need to undertake significant analysis of the format of, and content within, the data to understand what can potentially be obtained from it. The unstructured data will certainly need lots of processing power, or structuring, to release that value. That in turn means you will need to fund potentially complicated, protracted and expensive development costs.

The benefits of Big Data could therefore be significant on the one hand, but the analysis may be expensive and longer term on the other. Not the easiest business case for any governing body to sign off in today's world.

Robotic process automation and autonomous systems

Robotic process automation (RPA) is, just for once, a term that says it all. It is the use of digital technology and processing power (robots) to automate tasks (processes). Not all tasks will be suitable. They need to be repeatable and standardised, with well-defined and consistently formatted content. Automation involves machines (including computers) that are able to complete repetitive tasks at a fraction of the time it would take humans and to do so with greater accuracy, typically with less risk to humans, and without complaint.

Some robots are getting more and more like humans. When you are chatting to a customer service representative on a website it is more and more likely that "Jenny" is a robot and not a human, even if you don't realise it. And robots with cute or babyish characteristics (large eyes, round faces, small bodies) are being developed for customer service in hotels and public places and even as companions for the elderly.

There are a number of benefits from RPA and these include:

- **Reduced costs**. Costs can be reduced significantly by automating key processes. In addition, there may be a reduced need to outsource specialist tasks with a saving not just in money but also in reliability and in other risks (including security, privacy and resilience) that come from using third parties.
- **Better use of people**. Good talent is rare. And using it to undertake important but routine tasks that could be completed by a computer is wasteful for organisations, and frustrating for the talent. By freeing people up to do things that add greater value such as supporting decision making, organisations increase effectiveness.
- **Increased speed**. Processes can be completed more speedily. Even relatively complex tasks such as reviewing contracts or auditing companies can be conducted by computers. This reduces risk and opens more opportunities.
- **Improved quality**. Data processing quality can be improved with automation. There will be a reduction in error rates compared with human data input.
- **Higher availability**. With RPA you could be offering a continuous service, including at nights, weekends and during holidays.
- **Greater consistency**. Using a machine to undertake processes means that you can be sure that the process is always conducted in the same way, wherever. For instance, using RPAs for financial analysis across different business units would make it far easier to compare performance.
- **Better records**. If a machine undertakes a task it is simple to record when and where the task was undertaken and who caused the task to happen. It is much easier to ensure control and traceability with automated processes.

There are, however, several risks with RPA. It may stifle innovation. It may result in the retention of the status quo, merely automated, when better and different processes are available. And it may have an effect on your workforce creating worries (justified or otherwise) about job security, especially among people who do routine jobs such as recording transactions.

In addition, you will need to have the strength of mind to accept that, whilst the computer may be right a lot of the time, it can't always be right.

Software may have been programmed using the wrong data set. Or it may malfunction in certain circumstances. And there may simply be a mechanical breakdown such as a power outage. You will certainly need therefore to carefully identify how hand-offs from the digital world to the physical world will take place.

You will also have a need for constant monitoring and review, to ensure the behavioural and ethics expectations you have set are still being met. RPA won't be appropriate for every process, or for the whole of a process, or in every instance of the application of a process. An understanding of when the intervention of a human is needed in a process (e.g. to consider individual cases of hardship). And the resources to intervene promptly will also be needed if , for instance, the overall benefits of an automated process are not to become submerged a requirement to handle all exceptions.

The case for the adoption of autonomous systems is even clearer cut, with considerable savings in manpower and boosts to productivity to be obtained. Daily we see more examples of deployments and trials in public spaces, e.g. of robot vehicles.

But the arguments against such deployments continue, with a material level of concern about safety. At the time of writing, a tragic air accident had occurred with the death of over 150 people in an Ethiopian Airlines Boeing 737. Initial investigations, based on voice and data recorders, blamed the aeroplane's automated anti-stall system which appeared to force the nose of the aircraft down, despite all the pilots' efforts. In another example, in February 2019, the food delivery company Ocado's warehouse in Andover was burned to the ground, forcing the evacuation of hundreds of nearby homes and requiring firefighters to battle the fire for over 24 hours. Ocado said the fire started in a grid that allows robots to move around the facility. Counterclaims that it was started by a failure in the robotic system itself were later confirmed. This facility uses the robots to handle 65,000 customer orders per week and deliveries were, inevitably, seriously disrupted.

In a further example, a driver of a top of the range, new Jaguar reported that their car's speed sign recognition system was not very accurate on UK roads but on French roads was entirely unreliable. On several occasions, the autonomous system could not differentiate between 30 kph and 130 kph and often simply gave up. This, in turn, meant that the related function of automatic speed limitation failed, requiring driver intervention to avoid experiencing either frontal or rear-end collisions.

Artificial intelligence

Artificial intelligence is the ability of a machine to "learn" from past decisions and to then make its own decisions, thereby training itself to carry out a particular task better and better. It is often more correctly referred to as *machine learning*, which itself is a subset of artificial intelligence, along with computer vision, speech and language recognition, and pattern recognition. These are all types of narrow AI, which is different from general AI, a concept more similar to human brains and at the moment only seen in science fiction books and films.

A machine learns by modifying the importance it gives to the data it is using, based on the results of using that data. In this way, the output from the machine's decisions get ever closer to what is desired, at which point the machine will have learned how to carry out the task. In the financial services industry, machine learning could be used to judge whether a particular type of customer is worthy of being offered credit: if the learning is successful the rate of bad debts should decrease over time as the machine learns to make better decisions based on what types of customer have defaulted in the past. If you want to use personal data to make automated decisions such as these, it is as well to be aware of the restrictions that laws such as the EU's GDPR and the UK's Data Protection Act place on these activities.

In another example, somewhat challenging our views on what is appropriate and how easy (or not) it is to define and document the rules, the Estonian Ministry of Justice has asked the country's chief data officer to design a robot judge to take care of a backlog of small claims court disputes. The judge is supposed to analyse legal documents and other relevant information and come to a decision. Though a human judge will have an opportunity to revise those decisions, the project is a striking new application of artificial intelligence.

But rather less controversially perhaps, lift manufacturers Kone and ThyssenKrupp have both teamed up with IBM to use machine learning to predict operational problems such as worn lift parts before they cause trouble.

The biggest challenge to adopting AI, over and above the significant capital outlays required, will be the detailed process analysis that is required, for example, to understand how humans will process the information involved and how they will use it to make decisions. This is a matter of defining the rules by which the AI system will mimic human thinking, identifying what all the variables are that must be accounted for, what all the logic steps are, and finally what alternative outcomes are permissible at each point. Thereafter, you will need arrangements in place to constantly review and tune those rules.

Another very significant task with AI is to ensure the removal of bias in the rule sets and decision trees you use to control the AI. Humans *always* have biases. It is just that we often cannot spot them ourselves. You will need to undertake lots of reviews of those rule sets and decision trees, using different sets of eyes. These sets of eyes, with appropriate subject matter expertise and AI experience, that may well need to come from outside your organisation, will be hard to find and may be *very* expensive.

AMAZON ABANDONS SEXIST RECRUITMENT SOFTWARE

In late 2018 there were reports that Amazon, who had spent a number of years developing software to help with the identification of top innovators, had abandoned a machine learning software project because it was unintentionally but incorrigibly sexist. The tool was designed to give candidates for this assessment a rating between one and five stars depending on their CVs. Identifying the best applicants automatically would save executives a huge amount of time as well as potentially being more accurate and identifying much better talent overall.

Unfortunately, the company realised its new system was not rating those candidates in a gender-neutral way. This was because the software had been trained to evaluate candidates by identifying successful candidates over the past 10 years from the data within their CVs and validating that with other sources.

Unfortunately, the majority of the successful innovators over that period had been male, simply reflecting the past male-dominance of technology innovation that Amazon had been keen to destroy. This caused the AI software to value phrases like "Captain of the men's hockey team" but also to then dismiss phrases like "Captain of the women's hockey team".

Internet of Things

The Internet of Things (IoT) involves the connection of the Information Technology (IT) world with that of the Operational Technology (OT) world.

The connection of sensors and other controllers (many of which are already embedded in devices such as motors and which are increasingly cheap) to data networks enables you to operate, or simply monitor, very distant devices very efficiently, effectively and securely. This can deliver higher up time and thus lower units of production costs. For example, production machinery in your overseas plants may be linked to the internet in order for your engineers, here in your head office, to predict mechanical problems. And this same data, that is extracted as the machines are working, can also be used in your financial forecasting models.

Further efficiencies may come from the prospect of a smaller team of such engineers, (they won't be wasting time travelling between plants), and reduced capital outlays (they will need fewer vehicles and fewer sets of tools).

The Internet of Things is likely to be even more significant for your organisation if it works in conjunction with your use of other technologies such as RPA, Big Data and AI, VR (virtual reality) and augmented reality (AR) components to the solution. As nanotechnology, the ever-decreasing size of technology components, gains further ground and makes devices ever more intelligent, the viability of IoT itself will grow.

But there are risks. The primary risk is that of interoperability failures and risks to systems availability, caused by the interconnection of very different systems. In addition, the connection of sensors to networks means that, at a stroke, there are a host of new, vulnerable points through which cyber-attacks can be launched.

You may also find that the cultures of the integrated team that you will require will be at odds. IT operators have traditionally been more focussed upon integrity and confidentiality rather than upon availability. In contrast, engineering operators have been much more focussed on availability and, in their closed world of mechanical devices, little interested in issues related to confidentiality. You will need to specifically address this clash of cultures but, if you are successful, you will then have an unusually rounded and skilled team.

Wearables

Wearables are digital devices that can be attached to the body. Being highly portable, these devices allow immediate data input or access and thus mean that information is kept current. This can herald greater convenience and greater productivity. Keeping the hands free more of the time also means, of course, that those hands can be doing something else.

Clothing is one area where informationalisation (see Chapter 5) is growing fast: so-called smart clothing can contain sensors to monitor heart rate, breathing rate, acceleration, movement and the like. If you have a smartphone, you have a wearable, in as much as it is already counting your steps. In competitive team sports, safety can be built in with helmets that alert the coach if a player receives a blow to the skull, while the skills of the player and the team can be analysed with sensors that track a player's position on the field throughout a match. Smart clothing could potentially warn a worker of an imminent risk such as proximity to a noxious substance. And, connected to the Internet of Things, a wearable device might automatically summon an ambulance for the runner who has suffered a heart attack. While it is hard to think of any downsides for such technologies, the question you should perhaps ask is, do you really need to deploy this technology to do a job better? It may well be an expensive outlay of capital on a gimmick of little value.

3D printing

The mass personalisation described in Chapter 5 can be taken a significant step further with 3D printing (sometimes called additive manufacturing): the production of real 3D items as against 2D images of them.

Dentistry is one potential area of application, with 3D printing of braces, crowns and gum-shields becoming common. Sports helmets can be made to measure. And Swedish retailer Ikea has demonstrated a prototype video gamer's chair where the seat area can be custom produced, using 3D printing, to fit the user's bottom precisely.

This production technique can be very attractive in areas where there is high variability in production requirements. It can also be attractive in the production of consumables that are needed in places where space is at a premium or access for delivery is difficult. You could even think of it as

Just In Time (JIT) production, without the associated costs and time delays for shipping.

And there are other benefits. Highly complex items can be manufactured in one go, meaning that areas of weakness such as seams and joints are avoided. In addition, raw materials can be used that are sometimes lighter than traditional raw materials such as metals.

But there are real challenges with 3D printing. The production process can be complex, requiring detailed scanning, measurement, definition and modelling of the item – and that's if there is even a pattern to copy.

3D PRINTING AND CLASSIC CARS

One UK classic car company, faced by the lack of almost any spares availability for the vintage cars it repairs and restores, has turned to 3D printing. This has meant it can return a number of cars to the road, but at the cost of originality. This has mattered to some customers and not to others.

More importantly for them, the company has also faced a number of other, internal, hurdles. The stockroom space it hoped to release has become the 3D printing room. The stock-keeper it hoped to return to car engineering is now the 3D printing data analyst and operator. The capital equipment also dented the company's balance sheet as well as its profitability in the short term.

But as the owner of the business stated, it was do this or stop doing work on a whole series of vintage car makes.

Distributed ledgers and blockchain

Distributed ledger technology (DLT) is a way of recording transactions where the details of those transactions are held in various places at the same time. Because multiple copies of each record are held, it is difficult for any one person to change the details of a particular record without getting the agreement of all the other people who hold that record. It is intended as an open, and (globally) distributed, ledger that can record transactions between two parties efficiently and securely, i.e. in a verifiable and permanent way.

A blockchain is a type of DLT. In it, a growing list of records, called blocks, are linked. Each block contains an encrypted record of the previous block, a timestamp, and the relevant transaction data showing how the record has changed from the previous record.

Blockchains have been used for various cryptocurrencies, most famously Bitcoin. There have been real issues with considerable volatility of virtual currency units. For example, the value of Bitcoins was around $1,000 in January 2017, $15,000 in January 2018 and $3,500 in January 2019. It may be that this volatility ultimately puts the viability of cryptocurrencies in doubt. However, there are some genuine use cases for DLT beyond cryptocurrencies. These include:

- Creating a secure and credible record of financial transactions or contracts because, as they are "distributed", there is no single point of vulnerability as is the case with conventional records.
- Recording where money should be spent (e.g. by government on welfare or aid and by businesses on overseas investments) and how it was spent in reality when it has been distributed.
- Validating the origin of items such as food or jewellery.
- Increasing cyber security, for instance, by providing new ways of validating the identity of system users.

While these are all important objectives, it will always be necessary to ask whether distributed ledgers is a solution looking for a problem and whether there are simpler ways of achieving the desired ends.

Virtual reality and augmented reality

Virtual reality (VR) is a way of enabling someone to experience an environment different from the one they are in. Currently, a headset is generally used to provide visual and auditory information about the "virtual" environment. In some cases, this is supplemented by haptic information (touch, temperature and movement). For example, car designers can sit in a virtual representation of the car they have designed, to experience it and amend the details – e.g. of ill-fitting dashboard components.

In contrast, augmented reality (AR) presents individuals with additional, contextual information relevant to the tasks they are undertaking at the very point that they are performing it, e.g. lift manufacturer's technical notes presented to the lift engineer undertaking a lift service. This is analogous to head-up displays in aircraft and might, for example, be delivered via glasses – whether adapted versions of user's own or provided to those that don't otherwise need them. Some virtual reality headsets are capable of rendering augmented reality information at the same time.

The mixed reality spectrum

Virtual reality and augmented reality are not totally different concepts. They both fit into the mixed reality spectrum. This is a continuous spectrum of situations that lie between actual reality (the world we experience without any artificial aids) and virtual reality where we can be totally immersed in a digitally constructed world, one that generally is based on the real world.

The further along this spectrum one moves, the more difficult it is to maintain a link with actual reality, and therefore provide a credible experience. However, if a credible experience is delivered then the further one goes, the more immersed in the new reality the user is likely to be. At one end of the spectrum a pair of glasses (or a screen) may deliver additional information about the environment. At the other end, a user wearing a headset or even a complete haptic suit may think themselves in a different environment (Figure 14.1).

Virtual reality and augmented reality, especially when twinned with effective communications channels, both hold out the prospect of facilitating higher quality work outcomes from your personnel by the delivery and presentation of more immersive or more informed engagement in that

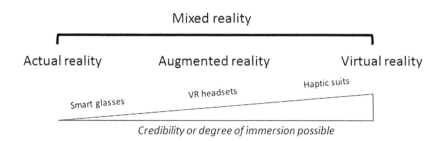

Figure 14.1 The mixed reality spectrum.

work. As well as use in live environments – which can, for example, be remote, inhospitable, spatially restricted or hazardous to humans – these technologies have proven to be effective in the training and development of employees and in the rehearsal of responses, e.g. to crisis scenarios.

But such technology has its limits. VR headsets can be fairly heavy, providing neck strain over time. Users also cannot communicate easily with others, meaning that there can be considerable time spent putting on and taking off the headsets.

But most importantly of all, not all your staff will be able to use them! Those staff with severe astigmatism, squints, macular degeneration, tunnel vision, etc. will find that the technology simply does not work for them. Clearly, you can be selective about who you get to work in virtual reality, but you risk hurting feelings, and claims of discrimination and associated actions.

Implants and brain-computer interfaces

At present, most interfaces between a human and a computer require some degree of human engagement – a finger hitting a keyboard, a voice command or a gesture. Brain-computer interfaces (BCI) increasingly seek to sidestep this requirement with the human mind directly interfacing with the computer. Implants are similar in that they don't require any physical action but differ in that they have a high degree of automation embedded instead of a human command.

At one end of the scale, it is at least sometimes possible to use the human voice (e.g. instructions to your car from voice and intonation synthesis) and gestures (offering potentials where keyboard access is constrained or there is high background noise meaning that voice commands won't work often enough).

At the other end of the scale, it is about emotions and thoughts being captured and instructions to devices. Science fiction? Not really. These techniques, while still primitive, are already being used to control drones, wheelchairs, prosthetic limbs and even video games.

While these ideas might sound like science fiction, they are starting to become reality. The benefits to people with paralysis are obvious and considerable medical research is happening in this area. But these advances are not just medical. The US Department of Defense has trialled BCI technology for drone pilots. And Boston-based start-up Neurable has created a video game called Awakening as a proof of concept for the thought control of video games. This is the technology that is likely to have huge implications in the near future.

While BCI technology isn't quite in the mainstream yet, digital implants very much are even if the percentage of the world population cover is "nano-scopic". Medical microchips designed for implanting in humans were approved by the U.S. Food and Drug Administration (FDA) as long ago as 2013. And in 2015 an office block in Sweden offered people who work there the change to be "chipped" as a way of gaining easy entry into the building.

Taking up such technologies raises both operational and ethical challenges. The adoption of voice and gesture commands ought to be straightforward enough but even here the technologies are still somewhat immature, user willingness to adopt them is still at a low level and the consistency of view as to what gestures and words mean what are yet to be properly codified with international standards.

The problems with implants and BCIs are far greater. Implants obviously pose massive ethical challenges, in a way that will dwarf the problems caused by smartphones and always-on working. How will you be able to assure those implanted that this is not mind-control, not mind-scanning, behaviour altering or sickness-inducing? And even if they go for it, what recompense are you going to offer for this corporeal invasion? And as for BCIs – who is to say that the direction of control won't just be from the human to the machine?

Is all this just snake oil?

We have described a few emerging technologies which may, or may not, deliver substantial opportunities to organisations in the years ahead. What is certain is that some people will inevitably overblow their potential because they are invested in their success or want to harness the hype. It's certainly important to treat the hype with caution. But that is not to say that these technologies are false, nor that they are not capable of making contributions to achieving your organisation's objectives. Most certainly will one day. And that certainty is driven by two things: Moore's Law and the compounding effect.

Moore's Law

Moore's Law asserts that, over time, digital technology gets cheaper and cheaper to manufacture and deploy. Compare, for example, the processing power of NASA's Apollo space missions with that of smartphones today. Also, in early 2019, the HP ZVR Virtual Reality system only cost a relatively affordable £5,338 for a 23.6-inch screen. Digital technology prices have been dropping continuously and will no doubt drop further. The real cost is often in using the technology however. For instance, the cost of the HP ZVR system will be in creating the images that can be shared on the screen.

What this suggests is that the business case for a technology that you rejected this month, as too expensive, could move into positive in a matter of months. You therefore need to have processes in place to re-evaluate technically good proposals on an ongoing basis. This will be especially true if you suspect that others, with a higher risk appetite, are likely to be piloting the technology.

The compounding effect

You should also take into consideration what might be called the compounding effect of combinations of these technologies. And this effect will apply to both the technologies positives (such as reduced costs of operation) and negatives (such as the likelihoods of incidents and scale of impact from those incidents, as well as the new complexities involved in ensuring resilience).

Certainly, some of the technologies discussed above will be reliant upon some of the others if they are to be fully effective. For example, RPA can really only be fully effective where an organisation has Big Data on hand, perhaps made possible by an Internet of Things approach.

THE COMPOUNDING EFFECT IN SMART METERING

The UK's Smart Metering programme is a seemingly innocuous technology programme to optimise both energy use and energy pricing. It is eventually to be a mandatory deployment and the cost will ultimately be met through customers' bills.

The deployment of Smart Meters in homes and businesses (essentially instances of the IoT), with user displays and command panels within the premises for control, connected to national networks to share consumption data with energy retailers and generators (who are planning to use AI to enhance performance and profitability) is set to increase the number of power consumption readings – per customer – by at least 17,500 times (Big Data). And software applications and services are already out there for remote – via smartphone – consumer management of their power consumption (more IoT).

Some of those energy retailers are using robots (RPA) as part of the customer service operations (e.g. chatbots that are used to engage with customers online). And at least one of these companies has been looking at the use of augmented and virtual reality tools (VR and AR) to identify, manage and fix faults in their smart metering networks.

But this compounding effect can also take you in directions that you don't immediately think of. We are not suggesting you start into areas close to the edge of what is legal, but even things like cannabis growing can be automated it seems.

WORLD'S FIRST AUTOMATED CANNABIS FARM

On 19 March 2019, Israeli company Seedo announced plans to build a fully automated, commercial-scale cannabis farm following Israel's recent approval of the export of medical marijuana.

The farm will consist of a series of stackable shipping container-sized units, each capable of supporting enough cannabis plants to yield 360 pounds of marijuana per year. Robotic arms will manage the cultivation, cameras will monitor the operation and machine learning software will be used to ensure optimal growing conditions. The start-up expects to yield four tons of dried cannabis during its first three years of operation. That's enough to generate a revenue of $24 million. So, it might not be long before there are automated cannabis farms cropping up wherever weed is legal.

Do the analysis and adopt with care

You can, and indeed you should, spend a proportion of your time dreaming of utopian futures – or perhaps imagining how to avoid dystopian ones. But do so objectively, being wary, and questioning, about the application of these new technologies.

One way to do this is to create an evaluation framework that might well include the consideration of:

- How does the technology really work now?
- What impact will its use have on your organisation's culture or the culture of other stakeholders? Will adopting it be in line with your organisation's declared ethical position and current digital governance framework or have you updated these (see Chapter 2)?
- Can you deliver the solution fully and does it help to evolve your state of digital adoption (see Chapter 3)?
- Does it offer any real opportunities to deliver improvements in your internal processes and will you be able to integrate it into your other operations (see Chapters 4, 5 and 6)?
- Can you manage the wider changes that will result, such as the impact it will have upon your workforce (see Chapter 9)?
- Can you use the technology safely?
- How would the use of this technology complicate (or simplify) your risk and compliance landscape and can you use the technology in such a way that will carry with it public and regulator confidence (see Chapter 10)?
- How will it impact your security, your customers' privacy and your organisation's resilience (see Chapters 11 to 13)?
- What future opportunities will this technology in turn open up (this chapter)?

In many ways, the adoption of these technologies underscores the need for the governance approaches and actions that are advocated in this book. We both wish you well with them.

DIGITAL GOVERNANCE GLOSSARY

3D printing   The process of creating a three-dimensional object from a digital model by laying down a series of layers of material (typically plastics or certain metals); it is a useful way of creating prototype machine parts quickly and cheaply; also called additive manufacturing

Access management   Managing who can access IT systems and the data contained on them as a way of ensuring data is kept confidential and integrious (a clumsy word meaning "having integrity")

Additive manufacturing   See 3D printing

Affective computing   Artificial emotion intelligence; a computer system that can recognise, respond to and simulate human emotions (sometimes called emotion AI)

Algorithm   A set of rules followed by a computer that are designed to enable it to undertake certain tasks such as making decisions

Artificial intelligence (AI)   The ability of computers to perform tasks that would normally need humans; AI involves intensive data processing (qv) and pattern detection that allows logical decisions to be made in ways that are often faster and better than humans; it consists of several elements including machine learning (qv) and computer vision

Augmented reality   The imposition of computer-generated data or imagery onto the visual field of a human providing an enhanced view of the world

Automation   The ability of a machine to perform a task repetitively and accurately, generally replacing the need for a human to perform the task

BCI   Brain-computer interface; an interface between a human's brain and a computer, either using a cable or, rarely, using brain waves, to allow a human to control a machine through thought rather than physical action

Behavioural targeting   Behavioural targeting enables you to show internet adverts to those unfortunate people who visited your site or searched for a product you sell, long after they have bought the garden shed, Greek holiday or leather belt they originally looked at

Big Data   Very large sets of data that may contain structured data (qv) and unstructured data (qv), the analysis of which can provide additional valuable insight to organisations

Biometrics   The use of physical or behavioural characteristics such as fingerprints or gait to identify individuals

Blockchain   A type of distributed ledger technology or DLT (qv); blockchain underpins cryptocurrencies like Bitcoin but has other uses too including ensuring the provenance of items such as gems, and cyber security

Bot   A software robot; a piece of software designed to interact with computers or humans autonomously

BYOD   Bring Your Own Device. The practice of workers using personally owned digital devices such as smartphones and tablets for work purposes such as answering emails from home or taking notes in work meetings; there are cyber security (qv) implications of allowing this.

Chatbot   A computer program (or "bot") designed to simulate human responses to questions asked online by a customer

CIA   Confidentiality. Integrity. Access. Referring to information quality, CIA are the tripod of requirements that cyber security systems must deliver

CISO   Chief Information Security Officer; the executive responsible for keeping corporate data and information safe from internal and external threats that may cause data breaches (qv) or other problems such as the inability to access data on a computer or use a computer system

Cloud computing   The storage of data and computer programmes remotely so they have to be accessed over the internet; sometimes defined as "using other people's computers" (like Amazon Web Services) although it is perfectly possible to own your own cloud computing infrastructure

Cloud computing   Use of a network of computers hosted remotely on the internet to store and process data, or make software available, rather than using a computer in your workplace

Cobot   A robot designed to work alongside humans safely, undertaking tasks that humans do slowly or poorly or which would be dangerous for humans to do

Consent   (In digital governance) the giving of permission for personal data to be processed. Consent may be explicit (denoted by an action) or assumed (denoted by inaction). Consent is one of the lawful reasons for the processing of personal data under GDPR (qv) but it is not the only reason

Cyber   Relating to computers and the internet

Cyber security   The protection of computers, computer systems and the data contained on them through the use of technology, business processes and human beings

Data breach   The unauthorised leaking of corporate data, either accidentally or deliberately

Data   Symbols such as numbers or words; often used to mean the same as information (qv)

Data classification   The process of sorting files and documents (data, qv and information, qv) into categories so that they can be managed more easily, for instance, for the purpose of securing the most important documents

Data controller   A person or organisation who defines the purpose of data processing (qv) and instructs it to happen

Data governance   The management of the processing of data used in an organisation, usually underpinned by a defined set of documented procedures

Data processing   The process of collecting, storing, sharing, manipulating or destroying data (qv)

Data processor   A person or organisation involved in data processing (qv)

Data subject   The individual whose personal data (qv) is being processed

Digital   Relating to computer technology including the hardware and the software that runs computers

Digital governance   The system of rules, practices and processes by which an organisation's acquisition and use of digital technology and digital information assets is controlled and monitored

Digital risk   Any business or human risk associated with a risk of damage to an organisation or an individual; data breaches (qv), reputational damage from social media, and health and safety compliance failures can all be examples of digital risk

Digital transformation   The process of radically changing an organisation through the use of digital technology to increase efficiency or provide new or greatly enhanced products and services. Not to be confused with creating a mobile app

Digital twin   A digital representation of a physical system, such as a machine or a city, designed to enable people to see in a risk-free manner how the components of the system interact and what would happen to the system under particular circumstances

Digitalisation   Digitisation (qv) with the intention of increasing the efficiency with which the information can be handled; an ugly word which we do not use in this book

Digitisation   The process of converting information into a digital form so that it can be manipulated by a computer, whether or not the intention is to increase efficiency

DLT   Distributed ledger technology; technology for creating a distributed, decentralised record or ledger that a set of people all have access to and control of so that if one of them wants to make a change all the others must agree

DPA   The UK's Data Protection Act 2019 which incorporates the wording of the EU's GDPR (qv)

DX   An abbreviation for Digital Transformation sometimes used in IT circles

Encryption   The process of converting data or information into a code so that it cannot be easily read by someone without the key to the code

Event   An action a computer takes as a result of an action taken by a computer user; every time a key is pressed an event happens

Fold (above/below the)   The fold in digital advertising means the bottom of the screen; anything "below the fold" cannot be seen unless the user scrolls down to it. Ads placed below the fold might not get seen as much as those above the fold but they often have higher engagement rates, presumably because users who bother to scroll are interested in the content of the page

GDPR   The EU's General Data Protection Regulation, a regulation which protects the personal data of people in the EU or of anyone globally whose personal data is processed by an organisation in the EU

Governing body   The people who define the purpose of an organisation policy, draw up the rules of how it should operate and monitor adherence to those rules by Top Management (qv)

Green IT IT systems that take account of ecological concerns; as computers use a lot of energy both for their operations and to cool them, ecological concerns can be significant

Haptic The perception of objects through touch, motion and proprioception (the position of the body); a haptic suit is a computer interface that takes signals from the position of a human's body (for instance in a virtual game of tennis) or transmits signals

ICO Information Commissioner's Office; the UK government agency responsible for regulating privacy

Incident In cyber security a computing event (qv) that has an impact (usually negative)

Information Data (qv) combined into structures (such as sentences) that have meaning to humans

Information lifecycle The stages information goes through: collection/created, retained, stored, retrieved, communicated, used and destroyed

Information quality The quality of the content of information systems; ideally information will be accurate, up-to-date, complete and unadulterated

Information security The protection of corporate and private information, whether held on computers or stored elsewhere, e.g. in paper ledgers

Internet of Things (IoT) The use of the internet to connect machines together, rather than people. For instance, a sensor in a field of crops could collect information on soil condition, pests or the weather and transmit it to a robotic agricultural machine that use that information to alter the way it moves across the field or the chemicals it sprays on the field

IT governance The processes that enable appropriate investments in information technology to be made and for that technology to be used appropriately, efficiently, safely and cost effectively

Knowledge A human's experience of the meaning of a particular piece of Information (qv)

Log A record of computer events (qv)

Machine learning The ability of computers to improve the decisions they make based on the results of previous decisions they made; one element of artificial intelligence (qv)

Maturity model A way of measuring the quality of an organisation's processes and delivery in a particular discipline (such as digital governance) used with the aim of enabling continuous improvement in that discipline

Metadata   Data about data; the "properties" of the document such as author, last modified by, title and date of creation are examples of metadata

Mixed reality   A spectrum where the real world and the virtual world are mixed in greater or lesser degrees so that at one end of the spectrum a human is given some additional information about the environment they are in while at the other end the human can become mentally immersed in a computer-generated environment

OCR   Optical character recognition, the ability of a computer to recognise typed and even handwritten text on a paper document and turn it into digitised text; can save a large amount of time. But it isn't always 100% accurate as documents with complex layouts, unfamiliar fonts, or smudged, crumpled and torn surfaces may not be read accurately. Trusting its accuracy without having it reviewed by a human can therefore be a mistake.

PCI DSS   The Payment Card Industry Data Security Standard; a worldwide standard set up to enable retailers and other companies that process credit card data to do so securely

Personal data   Data about an individual which can be used to identify them; a name like "John Smith" may not be personal data, but "John Smith, 21 Acacia Avenue, Happy Town" would be

PIA   Privacy impact assessment; a procedure to test risks to personal data during a particular process, as a way of making those processes more secure; PIAs are mandated under certain circumstances by the GDPR (qv)

PII   Personally identifiable information (or data); another term for personal data (qv)

Processing   Data processing; the process of collecting, storing, sharing, manipulating or destroying data (qv)

Quantum computing   A developing computer technology that relies on quantum physics rather than the 1s and 0s of digital computing

Robot   A machine designed to perform one or several tasks with precision and speed; robots differ from automated machines in that they can be programmed to change the way they operate

RPA   Robotic process automation; technology that enables computer software or robotic machinery to be programmed to emulate a human worker to eliminate tedious or dangerous tasks and allow them to focus on higher-value tasks

SAR   Subject access request; the right of data subjects (qv) to demand the information an organisation holds about them; a right under the GDPR (qv)

Search engine advertising   The process of paying for exposure on the results pages of search engines like Google in order to attract people to a website

Sensitive data   A common term for the "Special Categories of Data" outlined in the GDPR (qv)

Server   A computer that provides data to other computers

Shadow IT   The use of locally-grown, operated and maintained software and IT systems outside the official set of IT sanctioned by the organisation

Standard   A document that details an agreed set of best practice rules; standards provide specifications that can be used to ensure that digital products, processes and services are consistently fit for their purpose

Structured data   Data that has been formatted or structured in a way that makes the data easy to sort and process; an example might be the names, addresses and birthdates of all your customers

Tag   A piece of code on a website that collects data about the behaviour of visitors to the website

Top management   (In this book) the senior executives in an organisation who report directly to the governing body (qv)

TRL   Technology Readiness Level; a way of estimating the technical maturity and how near it is to commercial roll out of software or hardware

Turing test   A test to see whether a computer can imitate human behaviour (such as speech or writing) so well that a human observer mistakes the computer for a human

Unstructured data   Data (qv) that is not organised in a predefined way; an example might be a collection of your social media posts

UX   User experience; a term for the way that an end user of a computer system or programme experiences it in terms of the information they receive and the actions they need to undertake to interact with a system or programme; organisations should strive for a good UX but sometimes forget this

Virtual   Not physical, something that exists only in the digital domain

Virtual reality   A three-dimensional computer-generated simulation of an environment that allows humans to interact with it, thus creating a feeling of reality

INDEX

For Product Safety Concerns and Information please contact our EU
representative GPSR@taylorandfrancis.com
Taylor & Francis Verlag GmbH, Kaufingerstraße 24, 80331 München, Germany

www.ingramcontent.com/pod-product-compliance
Ingram Content Group UK Ltd.
Pitfield, Milton Keynes, MK11 3LW, UK
UKHW021118180425
457613UK00005B/136